David Hardie

Notes on Some of the More Common Diseases in Queensland in Relation to Atmospheric Conditions.

1887-1891

David Hardie

Notes on Some of the More Common Diseases in Queensland in Relation to Atmospheric Conditions. 1887-1891

ISBN/EAN: 9783337327439

Printed in Europe, USA, Canada, Australia, Japan

Cover: Foto ©berggeist007 / pixelio.de

More available books at **www.hansebooks.com**

NOTES

ON SOME OF THE MORE COMMON

DISEASES IN QUEENSLAND

IN RELATION TO

ATMOSPHERIC CONDITIONS.

1887–1891.

ILLUSTRATED BY CHARTS PREPARED FROM INFORMATION SUPPLIED BY THE REGISTRAR-
GENERAL'S DEPARTMENT AND CHIEF WEATHER BUREAU, BRISBANE, INCLUDING
RETURNS FROM THE HOSPITALS OF THE COLONY.

By

DAVID HARDIE, M.D.,

HON. PHYSICIAN, HOSPITAL FOR SICK CHILDREN AND LADY BOWEN MATERNITY
HOSPITAL, BRISBANE.

BRISBANE:
JAMES C. BEAL, GOVERNMENT PRINTER, WILLIAM STREET.
1893.

CONTENTS.

	PAGE.
Introduction	7
Diphtheria and Croup	11
Whooping-Cough	24
Phthisis	29
Pneumonia	36
Bronchitis	47
Respiratory Diseases	51
Diseases of the Circulatory, Nervous, and Urinary Systems	62
Diarrhœa and Dysentery	63
Gastro-Intestinal Diseases	75
Typhoid Fever	89
Malaria	104
Total and Comparative Mortality	115
General Summary	121
Appendix—The Flood of 1893	127

PREFACE.

In the following pages I have endeavoured to supply the missing link between the Mortality Statistics and Meteorological Observations in Queensland, for the quinquennial period 1887-91. From time to time we have received valuable reports from both the Registrar-General's Department and Chief Weather Bureau, Brisbane, but as yet no systematic attempt has been made to connect the one with the other. The subject is one which, I am convinced, deserves more attention than it has hitherto received, not only in Australia but in many parts of the Old World.

My main object has been to establish, on a more or less certain basis, the connection between the condition of the weather as represented by the various meteorological readings, with a few of the more common diseases. If this once be attained, a reliable forecast of the weather, such as our able meteorologist regularly provides, might at the same time supply us with a tolerably fair means of estimating the class of disease that may be expected, and directly or indirectly associated therewith. Should seasonal forecasts be also given, as proposed by the Postmaster-General, their importance in this connection can hardly be over-estimated. Lauder Brunton says—[1] "We may well fancy that the day is not far distant when warnings will be published in the newspapers, not only to seamen of approaching storms, but to invalids and people in general of the meteorological changes which will induce pain in some and nervous excitability in others, with perhaps an added hint that extra flannel should be worn by the former, and bromide of potassium or some other nervine sedative taken freely by the latter."

It is well to bear in mind that the conclusions arrived at, based as they are on mortality statistics, are only approximately correct as far as the condition of the atmosphere bears on the *causation* of disease, and that this can be but fairly ascertained even when supplemented by returns from the hospitals of the colony. In order to determine this

---

[1] "British Medical Journal," Jan. 2, 1892. "Lancet," Jan. 2, 1892.

B

more accurately, it would be necessary to have all cases of illness occurring within certain areas reported regularly to the State authorities, as is done in Norway and Sweden. Perhaps the time will arrive when the public will see the advisability of insisting not only upon the compulsory notification of all infectious diseases, but upon the weekly or monthly *registration of sickness*, in the interests of preventive medicine as well as for purposes of scientific investigation.

I must record my obligations to the Chief Secretary and the Colonial Secretary for the support they have given me in the matter, and for the liberality of the Government in authorising the necessary expenditure ; to the Under Colonial Secretary and the Assistant Under Colonial Secretary, to the Registrar-General, and to the Government Meteorologist, for their courtesy in supplying me with all available information and for generally assisting me ; and to the photo-lithographic departments for kindly reproducing the charts illustrating the paper.

The following Minute by the Colonial Secretary on my request for statistical information, well illustrates the liberal and scientific spirit of the Government, and deserves special acknowledgment :—
" Let the work be carried out as originally suggested. The investigations are in the cause of progress and, I feel sure, will justify the outlay." I can only hope that such is the case. Having had comparatively little to guide me in previous similar investigations, the remarks made may in some cases be unjustifiable, but with all my shortcomings in the matter, I may fairly say that the attempt has at least been an honest one to ascertain the truth.

The Notes on Typhoid Fever for Queensland as a whole were read before the Health Section of the Intercolonial Medical Congress held in Sydney in September last, and those on Diphtheria in Brisbane and Moreton District, before the Queensland Medical Society in October, and I trust that the publication of the whole, incomplete as it is, may induce others more competent than I am, to throw some fresh light on this uninviting though most interesting subject.

Wickham Terrace,
    Brisbane, February, 1893.

IT is the common experience of all of us that every now and then we have a run of some special class of disease. At one time we may have under our care an unusual number of enteric diseases, at another time laryngeal diseases, and at another time acute diseases of the lungs. In some instances we are unable to account for this particular rush, whereas in others, and with a certain amount of exactitude, we put it down to a more definite cause. Should we have a great number of cases of diarrhœa in November, December, and January, we think it is due to the excessive heat of summer, and if on the other hand we have an epidemic of pneumonia in the winter months, we attribute it to the cold and dry westerly winds that usually prevail at that time of the year.

It may be said in the case of alimentary diseases, especially in children, that the matter of diet is of more importance in point of causation than any seasonal atmospheric condition. We must bear in mind, however, that in general our diet is pretty much the same from one season to another, and almost certainly the same from one year to another; that in the former case if there is any difference it is one in the right direction and calculated to act rather as a preventive than an increased causative agent, and that in any case any change made is not sufficient to account either for the large percentage of cases of gastritis, enteritis, and dysentery met with during the hot months of late spring or early summer, or for the difference in mortality between one year and another. It is not impossible—it is rather probable—that the condition of the atmosphere in summer may be such as to effect some important changes in our food, and consequently the latter may be the *medium* through which some of these seasonal diseases are contracted. Although, however, our food may thus be the immediate cause of some diseases, the condition of the air has been the direct means of the change, and so has indirectly played the primary and essential part.

The same applies, and perhaps in even a more direct manner, to typhoid fever. Here we have a disease that is universally, and with truth, believed to be due to what is called an insanitary condition of our towns, villages, or districts as the case may be ; the specific germs being introduced either with our food or water supply through the stomach or from some foul-smelling sewer or cesspit through the lungs. But these are conditions—if not affected by some other agent—that would certainly remain with us with very little variation all the year round. Our water supply comes from the same source, and our food continues absolutely the same from one year to another. We have the same earth-closet, sewerage, or cesspit system, and the same nature of soil all round us. If during the summer there is any noxious effluvium arising from some drain, or rubbish heap, or cesspit, that does not exist in winter, it can hardly be due primarily to the habits of the occupants of the town or district, for here again we have conditions—whatever these may be—characteristic of the individual all the year round, and that probably never change during a whole lifetime. A person whose backyard is untidy or uncleanly in summer does not usually show more intelligent interest in matters pertaining to sanitation during the winter months, or, if so, the difference is not sufficient to account for the prevalence of typhoid fever at one time more than at another. Should he luckily escape the disease in winter, it surely is not because of extra care or good management on his part, but rather because nature has not supplied him with atmospheric conditions, either in temperature or rainfall, or perhaps both, favourable to the propagation of those bacilli that are known to be the immediate cause of typhoid fever. It is quite possible that his constitution may be more robust during the winter months, and therefore more able to successfully resist the influence of these germs; but here again we have more evidence in the same direction, for if there be increased resistive power in winter it must be due to the stimulating properties of the weather, and therefore to difference in atmospheric conditions between the various seasons of the year. So also in the case of phthisis. It is certain that conditions favourable to outdoor exercise—dry, bracing, and pure air, and plenty of sunshine—act a more important part, not only in the treatment but prevention of this disease, than all the active germicides we know of.

For many years the question of climate in the treatment and prevention of disease, notably chest affections, has created a great amount of interest, and several good authors, in almost every country, have given us the benefit of their experience and study in this particular direction. At the same time the very important matter of geographical pathology has been fully examined and has developed from an investigation into the distribution of disease in localised areas such as that made by Haviland, in 1875, on cancer, phthisis, and heart disease in England and Wales, until the question has become of world-wide interest, as evidenced by the work of Hirsch,[1] Felkin,[2] and Davidson,[3] and now embraces, in a systematic manner, both hemispheres. In many cases certain meteorological readings are given, chiefly regarding temperature and rainfall, as bearing on the cure, causation, and distribution of disease, but on account of differences in latitude and longitude, in soil, and physical conformation of the country between one place and another, it cannot be said that these observations in themselves, and with any certainty, have the same relation to health and disease as they may appear to possess. We know that although the meteorological observations of one district may be not unlike those of another it does not follow that we may meet with similar diseases, nor in the matter of treatment that the effects on disease are the same in both places. Neither does it follow that because the atmospheric conditions are different between one place and another they may not both present the same diseases, and be equally effectual in mitigating the symptoms of and perhaps in curing the disease. No

---

[1] "Geographical and Historical Pathology."
[2] "Geographical Distribution of Tropical Diseases."
[3] "Geographical Pathology."

matter, then, how carefully described by authors these meteorological readings may be, they can give us at best but an approximate estimate of their real value *per se* in relation to disease.

In forming a more accurate estimate of this connection between the two it will be necessary to compare the meteorological observations and mortality of one particular town or district, not with those in another part of the world where the configuration of the country may be very different, but with those obtainable in the same place — where local conditions are a constant quantity—from one month to another, one year to another, or perhaps for longer periods of time. Even here the relation shown can at present be but an approximate one, for, with the exception of temperature and rainfall, the remaining elements that together with these go to constitute what is ordinarily called the " weather " have not received by workers in this particular line of study the attention they possibly deserve. Consequently many mistakes will be made before an absolutely accurate idea of these matters can be formed, and before we can safely predict that certain diseases may possibly be expected in consequence of, or in connection with, certain conditions of the atmosphere ; but that it is the method most likely to give a faithful interpretation there can be little doubt.

With this object in view I have divided Queensland into eight large districts or divisions, and, through the liberality of our Government and the kind assistance rendered me by the Registrar-General's Department, obtained the mortality statistics for each separate division for the years 1887-91 inclusive.[1] At the same time I have received a record of meteorological observations for the same districts and for the same period of time, from Mr. Wragge, our enthusiastic meteorologist, who has taken great interest in the matter and assisted me in every possible manner. I have also obtained, through the Government, returns from nearly all the hospitals in Queensland, and have thus been able to compare these observations, not only with the death-rate but with the rate of admission to hospitals, for certain diseases for each division of the colony.[2]

It must be remembered that the results obtained are only approximately correct as far as meteorology bears on the *causation* of disease, for the illness might have been of several weeks' and in the case of chronic diseases of several months' duration, while hospital admissions, although showing a more direct connection, have the disadvantage of giving but a partial return of the diseases that prevail at any particular time or place. Moreover a still greater objection lies in the fact that in many inland districts in Queensland

---

[1] In the "Vital Statistics" published yearly the death-rate of Queensland as a whole and of Brisbane alone is given. In order to supply me with returns for the various divisions of the colony the Department was put to a considerable amount of trouble and expense.

[2] It is well to bear in mind that, while the "admission rate" applies to cases admitted into hospital, the " mortality " or "death-rate " includes not only those deaths that occur in hospital, but those also that take place in private practice outside the hospital, or, in other words, the general mortality of the district as given in the Registrar-General's reports.

the population is too small to supply, either from mortality or hospital statistics, a sufficiently large number of cases. In order to overcome this difficulty as far as possible, I have for Western Queensland taken the mean of the meteorological observations of two stations that are far apart such as Cloncurry and Hughenden, West Northern Queensland; Boulia and Blackall, West Central Queensland; Roma and Thargomindah, West Southern Queensland; and in one case the mean of three stations, Warwick, Toowoomba, and Dalby, as representing the Darling Downs. These are then made to represent, as they fairly do, the mortality and hospital returns of very much wider areas, and so give us a larger though perhaps less accurate basis to work upon. The same applies also to the other four divisions of the Colony, although here there is only one observing station for each district.

It would have been well had this comparison extended back at least ten years instead of five, but it is doubtful whether the meteorological observations recorded previous to 1887, when Mr. Wragge arrived in the Colony, were as absolutely correct as those we now possess. It is also well to bear in mind that the population of Queensland increased from 322,853 in 1886 to 393,718 in 1891, and that consequently the total number of deaths—other things being equal—should increase in proportion from one year to another.

The following are the observing stations and the districts they represent:—

Observing Station.	District Represented.
1. Brisbane	That portion of Moreton District within 30 miles of the coast lying between the parallels 27 and 28 deg. S.
2. Rockhampton ...	That portion of Port Curtis within 30 miles of the coast lying between the parallels 21 and 25 deg. S.
3. Cooktown ...	That portion of the Cook District near the coast lying between the parallels 14 and 17 deg. S.
4. Normanton ...	That portion of Burke inland from the Gulf of Carpentaria lying between 17 and 19 deg. lat. S. and 140 and 142 deg. long. E.
5. Cloncurry and Hughenden ...	That portion of Burke lying between 20 and 21 deg. lat. S. and between 140 and 145 deg. long. E.
6. Boulia and Blackall ...	That portion of North Gregory and Mitchell Districts lying between 22 and 25 deg. lat. S. and between 140 and 146 deg. long. E.
7. Roma and Thargomindah ...	That portion of the Warrego and Maranoa Districts lying between 26 and 28 deg. lat. S. and between 143 and 149 deg. long. E.
8. Warwick, Toowoomba, Dalby	That portion of the Darling Downs lying to the east of 151 deg. long. E.

DIPHTHERIA AND CROUP.

Diphtheria and Croup, Moreton District (Brisbane).—The mean monthly mortality curve of diphtheria for the five years 1887-91 for this district shows a gradual rise from January to its maximum for the year in June, followed by an equally regular and continuous fall to its minimum for the year in December : the mortality being low from October to February, high in May, June, and July, and medium during the other months of the year.[1] The combined croup and diphtheria curve follows a similar though somewhat more irregular course.

In comparing these curves with the various meteorological readings we find that it is essentially a cold-weather disease, and follows pretty closely in the inverse ratio the tracing of mean temperature, rainfall, number of rainy days, and amount of cloud ; that it fairly resembles the curve of the barometer and relative humidity; and that the mean and absolute range of temperature do not in themselves show any resemblance whatever.

One point here may be specially noted. It will be observed that the month of greatest mortality, June, occupies an intermediate place between the month of greatest humidity—May, and the month of lowest temperature—July. It would seem as if the death-rate increases as the temperature falls and the relative humidity rises, up till the month of May, that it increases still more in June in accordance with a further fall in temperature, but now that the humidity declines considerably, the mortality also falls, even though the temperature does not attain its minimum till July.

MORTALITY PER 100,000 POPULATION PER ANNUM for EACH MONTH of the YEAR.

Jan.	Feb.	March.	April.	May.	June.	July.	Aug.	Sept.	Oct.	Nov.	Dec.
24·0	21·6	38·4	50·4	68·4	93·6	70·8	48·0	36·0	21·6	15·6	12·0

RATIO PER CENT. for the FOUR QUARTERS of the YEAR.

—	Jan. to Mar.	April to June.	July to Sept.	Oct. to Dec.	Total.
Diphtheria	16·8	42·5	30·9	9·8	100
Croup	23·8	39·3	27·4	9·5	100

It will here be observed that the period of highest mortality from both diphtheria and croup is from April to June, and of lowest death-

---

[1] The above remarks apply to the mortality from diphtheria for the last five years in the Moreton district. According to statistics for the years 1880 to 1890 (11 years) for the city and suburbs of Brisbane alone, the month of highest death-rate from diphtheria alone or diphtheria and croup combined is not June but May ; the months of maximum rate for the year being April, May, June, and July.

rate from October to December, and that for every 10 deaths that
take place from diphtheria alone in the last quarter there are
17 the first, 43 the second, and 32 the third quarter of the year, and
from diphtheria and croup combined 21, 42, and 30 respectively.[1]

The following gives the mortality according to population for the
years as a whole :—

MORTALITY PER 100,000 POPULATION for the YEARS as a Whole.

—	1887.	1889.	1889.	1890.	1891.
Diphtheria	54·0	46·5	26·6	47·5	36·5
Croup	40·8	30·0	26·4	31·2	12·1

These figures will show that on an average for the last five years
for the district of Brisbane the combined death-rate from diphtheria
and croup is at the rate of 70·3 per 100,000 population, that there are
10 deaths from croup recorded by the Registrar-General for every 15
from diphtheria, and that in a general manner the years of highest
and lowest mortality for one are also the highest and lowest for the
other. It should be noted, however, that the very decided fall in the
death-rate from croup in 1891 is out of all proportion to the slight
decrease in the case of diphtheria.

If 1887 be taken as representative of a year of high mortality,
the following will show the meteorological observations for that year,
and, as a matter of comparison, those also representing the mean of
other years :—

MEAN MONTHLY METEOROLOGICAL OBSERVATIONS for the YEARS as a Whole.[2]

	1887.	Mean of 1888-91.
Barometer ...	30·036	30·080
Temperature ...	66·7°	68·7°
Mean range ...	16·8°	18·1°
Absolute range	34·1°	34·2°
Humidity ...	72%	69%
Rainfall ...	7·150	4·522
Rainy days ...	20	13
Cloud	5	5·2
Wind	10½	8½

---

[1] According to statistics for the years 1880-90 for the city and suburbs of Brisbane
alone (the above remarks apply to the district of Moreton including Brisbane), the
difference in mortality between one quarter and another is not so great—there being
for diphtheria alone and for diphtheria and croup combined 10 deaths the fourth
quarter for every 23 and 24 respectively the second quarter of the year. The ratio
for the various periods is as follows :—

	1st Quarter.	2nd Quarter.	3rd Quarter.	4th Quarter.
Diphtheria	10	23	15	10
Diphtheria and croup combined	12	24	20	10

[2] Barometer—Corrected and reduced to 32° and mean sea level.
Temperature—Mean temperature in the shade.
Mean range—Mean daily range of temperature.
Absolute range—Absolute monthly range of temperature.
Humidity—Relative humidity—saturation=100.
Rainfall—Amount of rainfall in inches per month.
Rainy days—Number of days on which rain falls per month.
Cloud—0 = sky cloudless ; 10 = sky perfectly overcast.
Wind—Velocity in miles per hour.

It will here be seen that a year as a whole of high mortality for both diphtheria and croup is coincident with a low temperature and low barometer, comparatively high humidity, high rainfall, and very large number of rainy days.

As regards this high death-rate from diphtheria for the year 1887 it will be found, in looking at the monthly curve, to take place chiefly in April and May, and that it falls considerably in June—the month that usually shows the highest mortality. In connection with the high rate during these two months, it may be mentioned that the mortality is at the rate of 121·5 per 100,000 per annum, while the average for other years is only 45·9, or not much more than one-third. Should 1890, where the mortality also rises high in May, be left out in forming a mean for other years, the rate is only 35·4 or *less* than one-third that of 1887. The following are the meteorological observations for the months of March, April, and May, 1887, together with those showing the mean of other years :—

MEAN MONTHLY METEOROLOGICAL OBSERVATIONS, MARCH, APRIL, and MAY.

					1887.		Mean of 1888-91.
Barometer	...	...	...	...	30·068	...	30·152
Temperature	...	...	...	...	68·1°	...	69·5°
Mean range	...	...	...	...	15°	...	15·4
Absolute range	...	...	...	...	29·6°	...	30·7
Humidity	...	...	...	...	79%	...	76%
Rainfall	...	...	...	...	6·158	...	5·031
Rainy days	...	...	...	...	24	...	16
Cloud	...	...	...	...	5·4	...	5·8
Wind	...	...	...	...	8	...	7

It will here be observed that the period of high mortality in 1887—April and May—is associated with low barometer and low mean temperature, high relative humidity and rainfall, and especially a large number of rainy days; and if 1890, where the death-rate also begins to rise early, were not included with the others to form a mean for 1888-91, the difference in relative humidity and rainfall would even be greater.[1]

It has already been stated that the mortality for the month of June, instead of increasing as is usual, begins to decline, and indeed is lower for this month than that of any other year.

---

[1] In connection with this high death-rate from diphtheria in the autumn months of 1887 and 1890, it may be of special interest to note that in January of the former year, and in March of the latter year, there were extensive floods in Brisbane whereby many of the low-lying parts of the town and suburbs were submerged in water, and subsequently covered, as the water receded, with foul-smelling mud and *débris*. Three months afterwards in the former case, and two months afterwards in the latter, the mortality from diphtheria rose very high. It should be mentioned that previous to the flood of 1890 the rainfall was unusually high in January and February, and this, together with the lower and therefore more favourable temperature of March as compared to that of January, 1887, may account for an interval of only two months in the one case, while there were three in the other. This would seem to indicate that there is a germination period in nature of two or three months' duration for the propagation and development of the germs of diphtheria.

During the past week (1st to 5th February, 1893) another flood—the most disastrous ever experienced by Europeans here—has visited Brisbane. It remains to be seen whether *it* also will be followed by an epidemic of diphtheria. If my observations are correct, this is highly probable during the months of April and May.

The reason of this fall from a meteorological point of view is at first sight not quite clear, for although the humidity is relatively low (65 per cent., while that of other years for the same month is 74 per cent.), and the rainfall equally low (·168, while that of other years is 1·536 inches)—conditions that seem to favour a low mortality—it should be noted that the temperature also is particularly low (56·3, while the mean of other years is 60·2 deg.)—a condition on the other hand that seems to be favourable towards a high mortality. It would appear either that a very low temperature in itself, and in the absence of a certain amount of relative humidity or rainfall, does not contribute towards a high death-rate, or that it positively retards the development of the disease. This latter suggestion is somewhat supported by a consideration of the mortality the following year—1888. Here we find that the rate is low the first few months of the year, and specially high during the two months of lowest temperature, June and July. In neither of these months does the temperature descend so rapidly as from May to June, 1887, and altogether it is comparatively high. It must be observed, however, that in July, 1889, the temperature descends still less, and yet the death-rate is particularly low, showing that mean temperature in itself during these months, whether high or low, is not invariably associated with high or low mortality.

In order to arrive at a reliable conclusion in the matter it is necessary that the atmospheric conditions of previous months be also taken into consideration. For the year 1888 there is no satisfactory explanation of the high mortality in June and July, unless it be the high rainfall in the month of February. In 1889 we shall find that the temperature has been particularly high from the beginning of the year, and that generally the conditions are not favourable to a high mortality from diphtheria. This also opens up the question as to whether there may not be a somewhat prolonged germinating stage in nature for the propagation of those germs that are associated with this disease, and that, in the absence of those conditions favourable to their development during the months of autumn, the mortality in winter will remain low, and *vice versâ.*

As far as the low death-rate for this year is concerned, it is well to bear in mind also that the atmospheric conditions are not favourable to the development of catarrhal and respiratory diseases generally, and as it is probable that these make one more susceptible to diphtheria and therefore act as predisposing causes, it follows that the death-rate from the latter will in consequence be somewhat reduced.

In 1890 we find that the death-rate is abnormally high in January and February (in continuation of a somewhat high rate the preceding month of December, 1889), and that this is coincident with a relatively high humidity and rainfall and low mean and absolute range of temperature, not only during these months, but for the last month or two of the preceding year. In January, February, and March the mean temperature also is comparatively low. For the next month or two the mortality remains moderately high, rises to a high maximum in May and June, being for these two months at the rate of 111·2 per 100,000 population per annum, while that of other years (leaving out 1887, where we have seen the mortality is also high), is 69·0, and then falls considerably in July.

It is important to note here that the temperature in July is not only lower by 5·2 deg. than that of June, but, with the exception of an almost equally low temperature in June, 1887, is lower than that of any other month under review. It has already been stated that the rapid decline in temperature in the presence of low rainfall and low relative humidity in June, 1887, is accompanied by a corresponding fall in mortality, and under similar conditions we find a similar fall in July, 1890. This fall cannot well be due *solely* in either case to the low relative humidity of the atmosphere or low rainfall, for these are comparatively low in 1888, and yet the mortality is high. It seems more probable, as has been suggested, that while a low temperature combined with high humidity generally tends to raise the death-rate from diphtheria, *an absolutely low temperature beyond a given point, if combined with low rainfall,* retards the development of those germs that are known to be the immediate cause of diphtheria, and that in consequence of this the mortality falls. It may be noted here that this is all the more suggestive when it is remembered that the death-rate from bronchitis and pneumonia is particularly high in the month of July, and as it is probable that the predisposing causes —colds and catarrhs—are equally prevalent, a diminution in the death-rate from diphtheria must necessarily mean that the exciting causes— the bacilli of diphtheria (Loeffler)—are all the less common or less virulent in their action.

With regard to the direction of the wind in its relation to diphtheria, it is impossible from mere monthly observations to come to any very reliable conclusion regarding the matter, on account of its variation from day to day, but the following remarks, based as they are upon reliable records, may be ventured upon.

For the month of April the mortality is very high in 1887, and here we find that the west wind was not observed once, and the south-west only twice. The south winds, on the other hand, were very prevalent, being observed 50 times, and forming 71 per cent. of the whole, while for the other years, 1888-91, they were on an average observed only 27 times and formed 42 per cent. of the whole.

For the month of May the death-rate is high in both 1887 and 1890, and the following figures will show the relative proportion of south, south-west, and west winds:—

		Times observed.				Per cent. of Whole.
South wind ...	{ 1887	...	...	... 32	..	54
	{ 1890	...	...	... 38	...	57
South-west wind	{ 1887	...	...	... 4	...	7
	{ 1890	...	...	... 2	...	3
West wind ...	{ 1887	...	...	... 18	...	30
	{ 1890	...	...	... 3	...	4

It would at first sight appear as if the south wind were intimately connected with a high mortality for the month of May also, but we find that in 1891 it was also prevalent, being observed 37 times and forming 51 per cent. of the whole, and yet the mortality was low. Neither can it be said that the west wind is invariably connected with a high death-rate for this month, for although it was observed in 1887 comparatively often it was noticed only 3 times in 1890 with an equally high mortality. It should be mentioned, however, that the west wind was observed very seldom for this month during the remaining three years, and that in these the death-rate from diphtheria was not high. The influence of the south-west wind in May is even more doubtful, for while it was observed on an average for 1887 and 1890 3 times, and formed 5 per cent. of the whole, it occurred in the remaining years 10 times and formed 16 per cent. of the whole.

For the month of June it is quite evident that the west winds are not intimately connected with a high mortality, for they were very much more frequently observed in 1887 than during any other year and yet we find the death-rate particularly low. Neither can it be said that the influence of the south wind is much more evident, for although it was observed in 1890 21 times against 4 times in 1889 (the mortality being high the former and low the latter year), it occurred 16 times in 1887 and formed 25 per cent. of the whole, the death-rate, as we have seen, being also low.

For the month of July the highest death-rate occurred in 1888, and it may here be noted that the south wind was very prevalent, being observed 25 times and forming 44 per cent. of the whole, while, if we take the average of others with low mortality, it was less common, being observed 18 times and forming 30 per cent. of the whole. The influence of the west wind is again very doubtful, for it was much more frequently observed in 1887 than 1888, in the proportion of 2 to 1, and yet in the former case the death-rate was less, in the proportion of 1 to 2. Again, while in 1888 the south-west wind was observed 7 times in the month of July, it was noticed 30 times in 1891, the mortality being particularly high in the one case and equally low in the other, and thus showing that if the south-west wind has any connection whatever the death-rate seems to fall as it increases in frequency.

Diphtheria and Croup—Queensland as a whole.—The mean monthly mortality curve of diphtheria alone or of diphtheria and croup combined for the whole of Queensland, for the years 1887-91, shows a gradual rise from January to its maximum in June, and then a fairly regular decline to its minimum for the year in November and December.[1]

The following gives the mortality according to population for each month of the year, from diphtheria and croup, and from both combined, together with the ratio per cent. for the four quarters of the year:—

MORTALITY PER 100,000 POPULATION PER ANNUM for EACH MONTH of the YEAR.

—	Jan.	Feb.	Mar.	April.	May.	June.	July.	Aug.	Sept.	Oct.	Nov.	Dec.
Diphtheria ...	16·2	25·2	27·6	32·4	54·0	66·0	46·8	38·4	26·4	25·4	16·8	15·6
Croup	19·2	15·2	26·4	36·0	36·0	34·8	28·8	19·2	22·8	12·0	13·2	7·2
Diphtheria and croup combined	35·4	40·4	54·0	68·4	90·0	100·8	75·6	57·6	49·2	37·4	30·0	22·8

[1] According to statistics for the years 1880-90, the mortality from diphtheria alone or diphtheria and croup combined is equally high for May and June, the months of maximum rate for the year being April, May, June, and July.

RATIO PER CENT. for FOUR PERIODS of the YEAR.

—	Jan. to Mar.	April to June.	July to Sept.	Oct. to Dec.	Total.
Diphtheria	18·4	39·9	29·5	12·2	100
Croup	22·7	39·5	25·5	12·3	100

It will be here observed that in the case of both diphtheria and croup the highest mortality takes place the second quarter, and the lowest the fourth quarter of the year, in the proportion of about 3 to 1.[1]

The following will show the death-rate according to population for the years as a whole, from diphtheria alone and from diphtheria and croup combined :—

MORTALITY PER 100,000 POPULATION for the YEARS as a Whole.[2]

—	1887.	1888.	1889.	1890.	1891.
Diphtheria	27·9	24·5	24·6	42·7	39·9
Diphtheria and croup ...	47·1	44·4	45·1	60·6	52·0

If the total death-rate from diphtheria be compared to that from croup, it will be found according to these figures that for the last five years there have been 14 deaths from the former recorded by Registrar-General for every 10 from the latter. Curiously enough, if we take the years 1887, 1888, and 1889 the mortality is upon the whole pretty nearly equal, whereas in 1890 the proportion of deaths from diphtheria to that from croup is nearly 2 to 1, and in 1891 the difference is greater still. In seeking for a possible cause for this relative increase of the former as compared to the latter during the last year or two, it is probable that professional opinion is now more in favour of classifying most cases of membranous inflammation of the throat as diphtheria, and that many of those that were formerly registered under the name of croup are now looked on as cases of diphtheria.[3]

---

[1] According to statistics for the twelve years 1880-91, there are for diphtheria alone or for diphtheria and croup combined 10 deaths the fourth quarter, for every 28 and 29 respectively the second quarter of the year. The following is the ratio for the four quarters :—

	1st Quarter.	2nd Quarter.	3rd Quarter.	4th Quarter.
Diphtheria	13	28	20	10
Diphtheria and croup combined ...	14	29	19	10

[2] According to these figures the average annual death-rate for the last five years per 100,000 population, for Queensland as a whole, from diphtheria alone is 31·9, and from diphtheria and croup combined 49·8. For the twelve years 1880-91, it is 25·5 and 52·5 respectively. In England the death-rate (1886-90) is 16·9 from diphtheria alone, and 29·4 from diphtheria and croup combined, showing that for every 10 deaths in England there are 17 in Queensland, according to population.

[3] This is supported by statistics for the years 1880-91. For these twelve years *in globo* the mortality from diphtheria is the same as from croup. When analysed, however, it will be found that from 1880 to 1884 the death-rate from the latter is nearly three times that from the former, while during the two years 1890-91 the proportion is reversed, so that the number of deaths from diphtheria is now nearly three times greater than from croup.

Taking 1890 as the year of highest mortality for diphtheria alone
or diphtheria and croup combined, the following are the meteorological observations connected therewith, together with those giving
the mean of other years for comparison : —

MEAN MONTHLY METEOROLOGICAL OBSERVATIONS for the YEARS as a Whole.

				1890.		Mean of Other Years.
Barometer	...	...	...	29·986	...	30·053
Temperature	...	...	...	71·7°	...	71·4°
Mean range	...	...	...	19·9°	...	20·9°
Absolute range	...	...	...	38·1°	...	39·5°
Humidity	...	...	...	67%	...	64%
Rainfall	...	...	...	4·333	...	3·078
Rainy days	...	...	...	9	...	7
Cloud	...	...	...	4·2	...	3·6

This will fairly show that the year of highest mortality is connected with low barometric pressure, somewhat low mean and
absolute range of temperature, and comparatively high humidity, rainfall, number of wet days, and high amount of cloud. It may be
noted that the temperature is relatively high, but this cannot in any
way be associated with the high mortality of that year, for it is higher
still (73·5 deg.) in 1889, and the death-rate, as has been seen, is for that
year very low.

It is highly probable, however, that the mean observations for
twelve months do not form as reliable a basis for comparison as those
we obtain from one month to another. In examining the monthly
curve it will be found for the summer months that the death-rate from
diphtheria is particularly high in December, 1889, and January and
February, 1890, and equally low during the same months of the
previous summer, the mortality in the former case being at the rate of
32·9, and in the latter only 6·6, per 100,000 population per annum.
The following table will give the meteorological readings for these two
periods : —

MEAN MONTHLY METEOROLOGICAL OBSERVATIONS, DECEMBER, JANUARY, and
FEBRUARY.

			1889-90.		1888-89.
Barometer	...	...	29·828	...	29·965
Temperature	...	...	79·9°	...	83·3°
Mean range	...	...	17·2°	...	23·7°
Absolute range	...	...	33·2°	...	44·2°
Humidity	...	...	66%	...	50%
Rainfall	...	...	9·152	...	2·366
Rainy days	...	...	15	...	7
Cloud	...	...	5·9	...	3·7

There is here a very distinct difference between the observations
of one year and those of the other. These show pretty conclusively
that during the warm months of summer the mortality from
diphtheria will be all the less, the higher the temperature, the higher
its mean and absolute range, and the higher the barometric pressure,
the lower the relative humidity, rainfall, number of rainy days, and
cloud; or, in other words, while a low summer temperature, combined

with great humidity of the atmosphere, increases the mortality from diphtheria during this period of the year, a high summer temperature and low rainfall seems to check or to be unfavourable towards its development. The same remarks apply also in the case of croup, though perhaps the difference here is not quite so decided.

It is possible, if any distinction can be drawn, that the moisture in the atmosphere is more intimately connected with a high mortality in summer than a low temperature, for during the period of high death-rate above noted, the temperature, though comparatively low in January and February, is relatively high in December. The humidity and rainfall, however, are exceptionally high, not only in January and February, but also the last few months of the previous year, and it is highly probable that these materially contribute towards the high mortality of summer 1890.

During the next three months of the year—March, April, and May—the death-rate from diphtheria is especially high in 1890 and equally low in 1888, being in the former case at the rate of 55·9 and in the latter 25·6 per 100,000 population per annum. The following table will give the difference in meteorological observations:—

MEAN MONTHLY METEOROLOGICAL OBSERVATIONS—MARCH, APRIL, and MAY.

	1888.		1890.
Barometer	30·118	...	30·021
Temperature	72·1°	...	72·1°
Mean range	15·6°	...	16·5
Absolute range	39·6°	...	32·1°
Humidity	62%	...	75%
Rainfall	1·016	...	5·654
Rainy days	4	...	11
Cloud	2·3	...	4·8

This would tend to show that a high mortality from diphtheria in the autumn months is associated with low barometric pressure and low absolute range of temperature, high relative humidity, rainfall, number of rainy days, and high amount of cloud; or, in other words, with conditions similar to those observed in connection with a high death-rate in summer. As regards the temperature, it does not appear to have the same intimate connection with the mortality as was observed in summer, or as will be observed in winter. We have seen that a high temperature in January and February apparently tends to reduce the mortality for that time, and we shall see that an absolutely low temperature in June or July seems to have the same effect, showing possibly that a medium temperature during these seasons is best suited for the development of diphtheria. It is not to be wondered at, therefore, as the temperature in the autumn months occupies an intermediate position between the temperature of summer and winter, that its connection with mortality in March, April, and May is inconstant and somewhat doubtful.

If we compare the death-rate for the month of June from
one year to another, we shall find that it is particularly high in
1891, medium for the preceding three years, and very low in 1887,
being at the rate of 100·6, 62·4, and 36·0 per 100,000 population
per annum respectively. The following table will show the respective
atmospheric conditions attending each :—

MEAN MONTHLY METEOROLOGICAL OBSERVATIONS for the MONTH of JUNE.

	1887.	Mean, 1888-90.	1891.
Barometer ...	... 30·110	... 30·130	30·073
Temperature...	... 56·8°	... 61·9°	58·4°
Mean range ...	.. 24·8°	... 20·7°	20·0°
Absolute range	... 40·2°	... 43·0°	39·6°
Humidity ...	... 70%	... 73%	73%
Rainfall ...	... ·582	... ·863	3·498
Rainy days ...	... 3	... 4	5
Cloud	... 3·3	... 3·2	3·5

It will here be noticed that the high death-rate for June, 1891, is
associated with low barometer and low mean and absolute range of
temperature, high rainfall and number of wet days, and high amount of
cloud. The low temperature for this year as compared to the mean of
the three preceding years must also be intimately connected with the
particularly high mortality in the former case, and yet the temperature
is lower still in June, 1887, the death-rate being at the same time lower
than that for the same month of any other year. It is well to bear in
mind, however, that this low temperature in 1887 is attended by very low
rainfall and few rainy days. This would appear to show either that a low
temperature in June does not increase the mortality in the absence of
a certain amount of rainfall, or, what is more probable, that while
a low temperature and moderate rainfall are indicative of high death-
rate (as shown in June, 1891), if the temperature descends too far and
below a given point (as in June, 1887), and especially if accompanied
by low rainfall, the developmental process is positively checked and in
consequence the mortality falls.

As regards the month of July the most striking feature from one
year to another is the very rapid drop in mortality from June to July,
1891, as will be seen by the following :—

MORTALITY PER 100,000 POPULATION PER ANNUM.

		1891.		Mean, 1887-90.
June ...	...	... 100·6	...	55·0
July ...	...	... 48·8	...	45·1
Total fall		51·8		9·9

The total fall in death-rate from June to July is 51 per cent. in
1891 and only 16 per cent in other years, or, in other words,
while in June, 1891, the mortality is almost double that of other
years, in July it is nearly equal. A comparison between the meteoro-

logical readings for each of these months may here be interesting, and perhaps give some explanation of this very decided and unusual fall.

TABLE showing the MEAN METEOROLOGICAL OBSERVATIONS for JUNE and JULY, and their RISE or FALL for the LATTER MONTH.

	1891.			MEAN, 1887-90.		
	June.	July.	July.	June.	July.	July.
Barometer	30·073	30·149	+ ·076	30·130	30·161	+ ·031
Temperature ...	58·4°	56·9°	-1·5°	60·6°	58·4°	-2·2°
Mean range ...	20·0°	23·4°	+3·4°	21·7°	22·3°	+ ·6°
Absolute range ...	39·6°	40·1°	+ ·5°	42·3°	40·6°	-1·7°
Humidity	73%	66%	-7%	72%	71%	-1%
Rainfall	3·498	·717	-2·781	1·043	1·551	+ ·508
Rainy days ...	5	3	-2	4	4	...
Cloud	3·5	2·6	- ·9	3·2	2·7	- ·5

These figures supply fairly conclusive evidence that the excessive fall in mortality from June to July, 1891, is coincident with a greater rise of barometric pressure (·076, while that of other years is ·031) ; a greater rise in mean range of temperature (3·4 deg., while that of other years is ·6 deg.) ; a greater rise in absolute range (·5 deg., there being an actual fall of 1·7 deg. in other years) ; a greater fall in relative humidity (7 per cent. against 1 per cent. for other years) ; a greater decline in rainfall (2·781, there being an actual rise of ·508 in other years) ; a greater fall in number of rainy days (2, there being no reduction in other years); and a greater fall in amount of cloud (·9, against ·5 for other years). As regards mean temperature it must be noted that it has not fallen so much as other years (1·5 deg. against 2·2 deg.) ; but it must be remembered that it is still comparatively low (56·9 deg.) ; and indeed, with the exception of the temperature for the month of June, 1887, which is equally low (56·8 deg.), it is lower than that of any other month under review. It has already been suggested that the low mortality during the winter months of 1887 is pretty intimately connected with the exceptionally low temperature in June for that year, and now that a relatively and absolutely low temperature in July, 1891, is accompanied by a rapid decline in mortality it may fairly be contended that the connection is not a mere coincidence, and that *if attended by dry weather, a very low temperature is not coincident with an increase, but rather with a decrease, in the death-rate from diphtheria.*[1] It is possible that the low rainfall and comparatively low relative humidity may both directly and indirectly help to reduce the mortality—directly by not supplying the moisture that is evidently necessary for the development of the diphtheria bacilli, and indirectly by allowing, on account of the greater diathermancy of the air, more loss of heat by night and greater increase of heat by day, and therefore a higher mean and absolute range of temperature.

---

[1] For the district of Moreton the death-rate also falls in June, 1887, and July, 1890, and apparently from similar causes.

C

Remarks on Diphtheria.

1. That though specially prevalent in the months of April, May, June, and July, it is endemic to some extent in Queensland during all seasons of the year.

2. That although the greatest mortality occurs in the autumn and winter months when the barometer is high, the lower the barometer during these months (1888, Brisbane, excepted) the higher generally will be the mortality.

3. That not only do the months of lowest temperature correspond generally to those of highest mortality, but the lower the temperature during the first few months of the year the higher the death-rate for that period, the sooner it rises, and the greater will be its maximum for the year; and conversely, the higher the temperature in January, February, and March the lower the mortality for that period, the longer it will be in rising, and possibly also the less will be its maximum for the year.

4. That a comparatively low temperature in autumn and winter is coincident with a high mortality.

5. That an absolutely low temperature, however, beyond a given point (58 deg.) in June or July, if attended by low rainfall, does not increase but seems rather to coincide with a decline in the mortality.

6. That a low mean and absolute range of temperature during the months of summer, and possibly also of autumn and winter, contribute towards a high mortality.

7. That the greater the relative humidity and the sooner it rises during the early months of the year the sooner the mortality rises for that year and the greater will be its maximum in autumn and winter (1888, Brisbane, excepted).

8. That although the months of lowest rainfall correspond closely to those of highest death-rate, and the months of highest rainfall to those of lowest death-rate, the greater the rainfall during the early months of the year the greater will be the mortality. That this high mortality will continue for a month or two even after the rainfall is reduced to perhaps below its normal amount.

9. That a high rainfall and high humidity in the latter part of the year not only seems to increase the death-rate for that period, but tends to bring about an early rise in the first months of the following year.

10. That a very high rainfall and high relative humidity in summer is associated with a specially high mortality the following autumn.[1]

11. That the greater the number of rainy days the greater will be the mortality (Brisbane excepted, 1888).

12. That the relation between the death-rate from diphtheria and amount of cloud is not so clearly defined.

13. That the influence of the wind in velocity and direction is (from monthly statistics) somewhat doubtful, although upon the whole a period of high mortality for the district of Brisbane is accompanied more by south than either west or south-west winds.

---

[1] *See* Appendix.

This variation of wind currents[1] much depends upon the course of the high and low pressure systems, and on the geographical position of their centres of energy. An anti-cyclonic or high-pressure system situated near Lake Eyre, and a V-shaped barometric depression showing steep gradients between New South Wales and New Zealand, give strong westerly winds at Brisbane, and as the high-pressure area makes easting so will the winds veer southerly, being the currents on the eastern or advancing side. These remarks refer especially to the winter months.

14. That atmospheric conditions favourable to the development of catarrhs and colds probably predispose to the development also of diphtheria.

15. That there is possibly a germination period *in nature*, of from two to three months' duration, for the development of the bacilli, or before the latter are sufficiently numerous or virulent to cause an epidemic of diphtheria.

16. From these observations it follows :—

(*a.*) That whatever contributes towards cold and dampness of the air during the months of autumn and winter causes an increase in the death-rate from diphtheria.

(*b.*) That as swamps and marshes increase the cold and dampness of any particular locality, independent to some extent of the condition of the weather, a thorough system of drainage is absolutely necessary to reduce the mortality from diphtheria for that district.[2]

---

[1] Wragge.

[2] These remarks apply to diphtheria only in so far as it is related to or influenced by the condition of the atmosphere, and in no way to the *medium* through which the bacilli are propagated. (*See* Appendix.)

WHOOPING-COUGH.

Whooping-Cough—Queensland as a whole.—The mean monthly mortality curve of whooping-cough for Queensland as a whole for the quinquennial period 1887-91 shows a gradual decline from January to June, a slight rise in July, a fall in August equal to that in June, followed by a moderate and uniform rise to December. This curve resembles very much that of mean temperature, rainfall, and cloud, and curiously enough is an exception to the ordinary rule of respiratory diseases, which generally exhibit a greater mortality during the cold months of the year. Here we have a disease that shows a higher mortality in summer than in winter, as will be seen from the following :—

MORTALITY PER 100,000 POPULATION PER ANNUM for EACH MONTH of the YEAR.

Jan.	Feb.	March.	April.	May.	June.	July.	Aug.	Sept.	Oct.	Nov.	Dec.
25·2	21·6	15·6	12·0	10·8	7·2	13·2	8·4	8·5	10·8	18·0	25·4

RATIO PER CENT. for FOUR PERIODS of the YEAR.

January to March.	April to June.	July to Sept.	October to Dec.	Total.
35·3	17·0	17·0	30·7	100

It will be noticed that the mortality is comparatively high in both the first and last quarters of the year, so much so that if we compare the death-rate from October to March with that from April to September we shall find it to be almost in the proportion of 2 to 1. If the death-rate were high only during the latter months of the year, it might be supposed that as the disease is often a protracted one it originated during the previous cold months of winter. According to the above figures, however, it appears to be even higher during the first quarter of the year, and altogether it may fairly be concluded that it is a disease rather of moist warm weather than of the cold dry season of winter.

The mortality for the years as a whole according to population is as follows :—

MORTALITY PER 100,000 POPULATION for the YEARS as a Whole.[1]

1887.	1888.	1889.	1890.	1891.
17·8	20·8	23·7	7·4	4·3

[1] The average annual death-rate for Queensland as a whole for the years 1887-91 is 14·6 per 100,000 population. This corresponds closely to the rate for the years 1880-91, which is found to be 14·1. In England the mortality (1886-90) is 44·36, or fully three times greater than in Queensland, in proportion to the population.

These figures show that the highest mortality occurs in 1888 and 1889, and the lowest in 1891. For convenience the meteorological observations for the years as a whole may be re-inserted:—

MEAN MONTHLY METEOROLOGICAL OBSERVATIONS for the YEARS as a Whole.

—	1887.	1888.	1889.	1890.	1891.
Barometer	30·016	30·074	30·055	29·986	30·031
Temperature	70·1°	71·7°	73·5°	71·7°	70·4°
Mean range	19·4°	21·6°	21·2°	19·9°	20·7°
Absolute range	35·4°	42·7°	42·0°	38·1°	37·8°
Humidity	71%	61%	63%	67%	63%
Rainfall	3·979	2·222	2·369	4·333	3·742
Rainy days	9	5	7	9	8
Cloud	3·9	2·9	3·7	4·2	4·1
Wind	11½	12½	12	12½	14½

It will be observed that the higher mortality of 1888 and 1889 is coincident with high barometer and high mean and absolute range of temperature, low relative humidity and rainfall, and low amount of cloud. In 1891 the atmospheric conditions are in some respects reversed, but not more so than we find in other years when the death-rate is comparatively high, showing that *for the years as a whole* there is no constant connection between meteorological readings and the death-rate from whooping-cough.

When reference is made, however, to the monthly curve, it will be found that in most cases the high mortality of one year in its early months is simply a continuation of a high death-rate during the last months of the previous year, and is possibly, therefore, associated with the atmospheric conditions of the year preceding. As an instance it may be noted that the period of high mortality in 1888 begins in November, 1887, is especially high in December and January, and continues moderately high till May. If we take the three months of highest mortality—November, December, and January—we shall find that the average death-rate during that time is at the rate of 49·1 per 100,000 population per annum, whereas the mean of other years for the same months is only 15·1. In order to ascertain if possible what connection this high mortality has with the condition of the atmosphere, it will be necessary to obtain the meteorological readings of the previous month or two, as it is probable the disease may have been of some duration.

MEAN MONTHLY METEOROLOGICAL OBSERVATIONS—SEPTEMBER, OCTOBER, and NOVEMBER.

	1887.		Mean of Other Years.
Barometer	30·031	...	30·039
Temperature	70·8°	...	74·1°
Mean range	23·3°	...	24·1°
Absolute range	42·4°	...	44·1°
Humidity	63%	...	55%
Rainfall	1·511	...	1·559
Rainy days	7	...	5
Cloud	3·5	...	3·0
Wind	12	...	13½

It will be observed that there are associated with the high mortality
of this period during the months of spring, 1887, a low temperature,
comparatively low mean daily and absolute range of temperature, high
relative humidity and number of wet days, and high amount of cloud ;
and if the comparison be continued for the summer months of 1888 we
shall find the same comparatively low temperature (78·9, while that
of other years is 80·5 deg.), together with a moderate rainfall and
fairly high relative humidity. It might at first sight appear as if
this low temperature does not quite coincide with what has already
been stated—namely, that whooping-cough shows its highest mortality
during the warm months of the year ; but although that may be so,
it does not follow that the higher the temperature is in summer the
greater will be the mortality. It would certainly appear, and it
is quite consistent to suppose, that whooping-cough requires a
moderately high temperature for its development, and that if the
latter rises beyond a certain point the death-rate falls. Moreover,
in this instance it is associated in the latter months of the
year 1887 with high relative humidity, large number of rainy days,
and high amount of cloud, and in the early months of the following
year with medium amount of rainfall and comparatively high relative
humidity, all of which support the assumption of a germ origin for this
disease, and requiring for purposes of propagation a moderately high
but not too high temperature, and a certain amount of moisture.

If this supposition be correct, then the *higher* the temperature,
relative humidity, rainfall, and amount of cloud in the winter months,
the greater should be the mortality for the time being and the succeeding few months. In 1889 we find that from June to November
there are fully twice as many deaths recorded as on an average for the
other four years, and the following are the meteorological observations
connected therewith :—

				1889.		Mean of Other Years.
Temperature	...	...	...	61·4°	...	59·7°
Humidity	...	...	...	73%	...	68%
Rainfall ...	...	...	...	1·540	...	1·323
Rainy days	...	...	...	5	...	4
Cloud	...	..	...	3·3	...	2·7

showing a distinctly higher temperature in connection with the
epidemic in the winter months of 1889, and from which we may
fairly conclude that the mortality from whooping-cough is all the
greater during the winter months and those immediately following, the *higher* the temperature and the greater the relative humidity
of the air. It would therefore appear that whooping-cough is favoured
in its development by a *high* temperature in winter and an equally *low*
temperature in summer, in combination with high amount of cloud and
a humid atmosphere during all seasons of the year.

Whooping-Cough—Brisbane District.—The mean monthly mortality curve of whooping-cough for this district is pretty much the

same as for the whole of Queensland, showing a maximum rate in the warmer and a minimum rate in the colder months of the year.

RATIO PER CENT. for FOUR PERIODS of the YEAR.

January to March.	April to June.	July to September.	October to Dec.	Total.
34·5	8·9	14·1	42·5	100

The following will give the death-rate according to population for the years as a whole:—

MORTALITY per 100,000 POPULATION for the YEARS as a Whole.

1887.	1888.	1889.	1890.	1891.
30·0	14·4	34·8	16·8	1·6

It will be seen that the mortality is specially high in 1887 and 1889, and that in 1891 it is reduced to a very low minimum.

In examining the monthly curve from one year to another we shall find that this high mortality in the former instances is due chiefly to a high maximum in November and December, 1887, and to a lower and earlier one from September to December, 1889. As regards these the following tables will give the meteorological observations in the one case for September, October, and November, 1887, and in the other for June, July, and August, 1889, with the mean of other years for purposes of comparison.

MEAN MONTHLY METEOROLOGICAL OBSERVATIONS.

	SEPT., OCT., AND NOVEMBER.		JUNE, JULY, AND AUGUST.	
	1887.	Mean of Other Years.	1889.	Mean of Other Years.
Barometer	30·034	30·077	30·121	30·131
Temperature	67·1°	69·4°	60·1°	58·8°
Mean range	20·0°	19·9°	17·3°	19·9°
Absolute range	37·3°	38·9°	36·6°	36·6°
Humidity	65%	61%	74%	71%
Rainfall	3·238	3·406	4·338	2·697
Rainy days	23	10	9	7
Cloud	4·5	4·9	4·6	3·5
Wind	12	8½	9½	8½

We have here further evidence, as observed for the whole of Queensland, that a high mortality is associated with a high winter and low spring or summer temperature combined in all cases with high relative humidity and large number of rainy days; and if the germ origin of disease applies to whooping-cough—which it very probably does, as shown by its contagious nature—the atmospheric conditions above noted, moderate temperature and high humidity, are highly favourable towards its development.

REMARKS ON WHOOPING-COUGH.

1. That, unlike respiratory diseases generally, it attains its maximum mortality during the warm and moist months of the year.

2. That a high temperature in winter is coincident with a high mortality for that period as well as for the succeeding months of spring.

3. That a low temperature during the latter months of the year is associated with a high mortality for that period as well as for the first few months of the following year.

4. That a moderate rainfall and high relative humidity are generally connected with high death-rate.

5. That the association of a high mortality with medium temperature for all seasons of the year, in combination with moderate rainfall and high relative humidity, supports the assumption of a germ origin for whooping-cough.

PHTHISIS.

Phthisis—Queensland as a whole.—The mean monthly mortality curve of phthisis for the whole of Queensland for the quinquennial period 1887-91, shows a depression in February, a minimum rise in March, and a maximum rise in August, so that altogether for the twelve months, the death-rate is low in February, high in March, July, August, and September, and medium in January, April, May, June, October, November, and December.

In comparing this curve with the various meteorological readings for the same period, we find in a general way that the period of highest mortality corresponds fairly to high barometric pressure and high mean and absolute range of temperature, low temperature, rainfall, and number of rainy days and low amount of cloud, and that the period of low mortality corresponds to atmospheric conditions pretty nearly the opposite of these.

With regard to the fall of mortality in February, it can hardly be looked on as the result of a continuously high temperature during the previous months of summer, nor to the slight fall in mean and absolute range of temperature, because the same conditions, even to a more marked extent, together with increase in humidity, rainfall, and amount of cloud, apply to January as compared to the last quarter of the previous year, and yet we find the mortality for these four months is nearly equal. It would appear rather that the mortality in February is lower, and that of January higher, than what the atmospheric conditions would lead us to expect. This might be due in the latter month not only to the addition of a few deaths that should naturally have taken place in November and December previous had the weather been less favourable, but to the enervating influence upon patients in the last stage of consumption, of a prolonged high temperature, and thus not only increasing the death-rate for that month to an undue extent but at the same time leaving all the fewer to die in the month of February. Weber says,[1] "In many weakly subjects we constantly observe increased energy of all functions, better appetite, and greater ease of muscular action during warm weather, but if exposed to it for a longer period the appetite fails, and the function of the digestive organs and of the nervous system become impaired."

In March the weather undergoes a fairly distinct change, and coincident therewith the mortality rapidly increases. After a few months of a continuously high temperature, the latter, together with the amount of cloud, falls, while the barometer rises, and the wind changes in its direction and increases in velocity.

For the next three months the death-rate is low, and under the following atmospheric conditions: a falling temperature, amount of

[1] Von Ziemssen's "Handbook of General Therapeutics."

rainfall, and cloud, a rising barometer and mean daily and absolute range of temperature, together with a rise in relative humidity to its maximum for the year. As the presence of some of these might lead us to expect a higher death-rate, it is possible that the distinct change in the weather in March has increased the mortality to an undue extent, leaving comparatively few to die during the next month or two, and so reducing the number of deaths for the time being.

In July the temperature is at its lowest for the year, and under a high barometer, high mean and absolute range of temperature, low rainfall and low amount of cloud, and falling humidity, the mortality now rises, and even in the presence of a slightly higher temperature continues to rise still higher to its maximum in August. In September and also in October, while the relative humidity declines and the mean and absolute range of temperature remain high, the mortality rapidly falls. This is probably due mainly to a decided rise in temperature, slight rise in amount of cloud, and fall in barometer, combined perhaps with more favourable winds. It is possible, also, that the high death-rate in the winter months necessarily brings a diminution in the mortality during the next month or two. The atmospheric conditions now become more favourable, the barometer and mean and absolute range of temperature fall, the temperature, relative humidity, rainfall, and amount of cloud rise, and for the next few months the mortality remains uniformly low, falling to its minimum for the year in the succeeding month of February.

The mortality from phthisis, according to population, on an average of five years for the various months of the year, and also the ratio per cent. of deaths for the four quarters of the year, may here be of interest:—

MORTALITY PER 100,000 POPULATION PER ANNUM for the VARIOUS MONTHS of the YEAR.

Jan.	Feb.	March.	April.	May.	June.	July.	Aug.	Sept.	Oct.	Nov.	Dec.
129·6	108·0	138·0	129·6	124·8	121·0	144·1	160·8	144·0	124·8	120·0	124·8

RATIO PER CENT. for FOUR PERIODS of the YEAR.

January to March.	April to June.	July to September.	October to December.	Total.
23·9	23·9	28·7	23·5	100

showing that, although the number of deaths is greater from July to September than in any other period of the year, the difference is not great, and certainly less from one quarter to another than that of any other disease under consideration.

PHTHISIS.   31

The following tables will give the relative mortality according to population and the meteorological observations from one year to another:—

MORTALITY PER 100,000 POPULATION for the YEARS as a Whole.[1]

1887.	1888.	1889.	1890.	1891.
130·8	140·1	128·6	135·7	127·3

MEAN MONTHLY METEOROLOGICAL OBSERVATIONS for the YEARS as a Whole.

—	1887.	1888.	1889.	1890.	1891.
Barometer	30·016	30·074	30·055	29·986	30·031
Temperature	70·1°	71·7°	73·5°	71·7°	70·4°
Mean range	19·4°	21·6°	21·2°	19·9°	20·7°
Absolute range	35·4°	42·7°	42·0°	38·1°	37·8°
Humidity	71%	61%	63%	67%	63%
Rainfall	3·979	2·222	2·369	4·333	3·712
Rainy days	9	5	7	9	8
Cloud	3·9	2·9	3·7	4·2	4·1
Wind	11½	12½	12	12½	14½

It will here be observed that although the mortality is somewhat higher in 1888 and 1890, it does not vary much from one year to another. In both these cases the temperature is medium and its absolute range high; in one the barometer is high and the relative humidity, rainfall, number of rainy days, and amount of cloud low, while in the other these are reversed. It may also be noted that while the year of lowest mortality—1891—is associated with low temperature and low absolute range of temperature, in 1889, with an almost equally low death-rate, the temperature and its range are unusually high, the relative humidity being low in both cases. It would seem, therefore, as if there is no constant connection between *years as a whole* of high or low mortality with certain meteorological observations, although in a general manner a year of low death-rate coincides with a low temperature and low absolute range of temperature, the rainfall and amount of cloud being at the same time fairly high.

In comparing the monthly curve from one year to another, one peculiarity observed is, that the death-rate in the autumn months of 1889 is comparatively high, attaining as it does its maximum

[1] According to these figures, the average annual mortality from phthisis for Queensland as a whole for the years 1887-91 is 132·5 per 100,000 population. It may be stated that for the twelve years 1880-91 it is slightly higher—namely, 152·3—showing that the death-rate for the last few years is not so high as it was in the early years of the last decade. In England, the death-rate from phthisis (1886-90) is 163·54 per 100,000 population, there being thus about 12 deaths in England to every 10 in Queensland according to population. This, however, does not give a fair estimate of the relative prevalence of the disease among Europeans here and at home, as in the former instance the mortality is increased by the addition of imported cases that die here, as well as by an unusually high death-rate among Polynesians.

for the year in March, while in the winter months it is comparatively low. The following tables will show the relative proportion of
deaths according to population for these two seasons, and also the
meteorological observations connected therewith :—

MORTALITY PER 100,000 POPULATION PER ANNUM.

MARCH, APRIL, AND MAY.		JUNE, JULY, AND AUGUST.	
1889.	Mean of Other Years.	1889.	Mean of Other Years.
153·6	125·6	128·1	146·1

MEAN MONTHLY METEOROLOGICAL OBSERVATIONS.

	MARCH, APRIL, AND MAY.		JUNE, JULY, AND AUGUST.	
	1889.	Mean of Other Years.	1889.	Mean of Other Years.
Barometer	29·965	29·858	30·147	30·146
Temperature	83·3°	79·4°	61·4°	59·7°
Mean range	23·7°	17·3°	20·3°	23·2°
Absolute range	44·2°	33·8°	41·8°	42·1°
Humidity	50%	67%	73%	68%
Rainfall	2·366	7·875	1·540	1·323
Rainy days	7	13	5	4
Cloud	3·7	5·4	3·3	2·7
Wind	11	12½	12	13

These figures show very distinctly that in the autumn months
a high mortality is connected with a high barometer, very *high*
temperature, and mean daily and absolute range of temperature,
combined with very low relative humidity, rainfall, number of wet
days, and low amount of cloud. In winter, on the other hand, a
high death-rate is associated with *low* temperature, while the other
atmospheric conditions are the same, though not so extreme, as
those observed in connection with a high death-rate in autumn.
It would appear, therefore, that a very high temperature during
the early months of the year is unfavourable, and increases the
death-rate for that period ; but if it keeps up in the winter
months above the average, the rate will then be reduced, not
only below the mean of other years, but even below that of the
previous months of autumn. It is possible, however, that a low temperature in winter is not unfavourable, provided the absolute range
of temperature is also low and the rainfall moderate, for in 1887 and
1891 the death-rate is low during the winter months in both cases,
the temperature and its range being also low. Conversely a high
temperature is not favourable in winter if its absolute range is high
and rainfall low, for in 1888 and 1890 the mortality is high while the
temperature is comparatively high, its range absolutely higher than
that of any other year, and the rainfall extremely low.

Phthisis—Brisbane District.—The mean monthly mortality curve of phthisis for this district is somewhat irregular, the death-rate being high in March, April, and September, low in May and October, and medium for the remaining months of the year. The following will give the monthly rate according to population and the ratio per cent. for four periods of the year :—

MORTALITY PER 100,000 POPULATION PER ANNUM for EACH MONTH of the YEAR.

Jan.	Feb.	March.	April.	May.	June.	July.	Aug.	Sept.	Oct.	Nov.	Dec.
137·40	131·28	169·68	155·40	115·08	143·40	147·48	147·48	173·76	123·24	139·20	138·36

RATIO PER CENT. for FOUR PERIODS of the YEAR.

January to March.	April to June.	July to September.	October to Dec.	Total.
25·4	24·1	27·2	23·3	100

These figures show that there is a fairly uniform rate from one period of the year to another, although it may be noted that the percentage of deaths is somewhat greater the first quarter and less the third than for the whole of Queensland.

Taking the years as a whole, the following will give the death-rate from one year to another, and, for comparison, the meteorological observations coincident therewith :—

MORTALITY PER 100,000 POPULATION for the YEARS as a Whole.

1887.	1888.	1889.	1890.	1891.
124·8	163·2	141·6	159·6	126·0

MEAN MONTHLY METEOROLOGICAL OBSERVATIONS for the YEARS as a Whole.

—	1887.	1888.	1889.	1890.	1891.
Barometer	30·036	30·128	30·077	30·039	30·071
Temperature	66·7°	68·6°	69·6°	68·7°	68·4°
Mean range	16·8°	19·1°	17·4°	17·5°	17·6°
Absolute range	34·1°	36·2°	35·0°	32·5°	32·7°
Humidity	72°/₀	66°/₀	69°/₀	72°/₀	69°/₀
Rainfall	7·150	2·960	4·115	5·160	4·030
Rainy days	20	13	12	14	12
Cloud	5	4·5	5·5	5·7	5·1

According to these figures there is no constant connection between a year as a whole of high or low death-rate from phthisis and meteorological readings, although, if we take 1887 as representative of a year of low mortality and 1888 as one of high mortality, it will be found that the former corresponds to low barometer, low temperature and low mean and absolute range of temperature, together with high relative humidity, rainfall, and cloud, and very large number of wet days, while the latter is associated with conditions the reverse of these.

In comparing the monthly tracing from one year to another, there is so little difference between one and the other, that it is impossible to state in exact terms to what extent the various meteorological conditions tend to raise or lower the mortality from phthisis for the district of Brisbane. It will be found, however, that in summer the lowest death-rate occurs in 1887 and 1889, that in autumn the death-rate is nearly equal each year, that in winter it is lowest in 1887 and highest in 1890, and that in spring it is lowest in 1891 and highest in 1888.

As regards the low mortality in the summer months of 1887 and 1889, it will be noticed that in the former year the barometer, temperature, and mean and absolute range of temperature are low, and the relative humidity and number of wet days and amount of cloud are high, while in the latter year the opposite conditions are observed. It might, therefore, appear as if really the death-rate from phthisis is quite independent of the condition of the weather during the summer months. If we examine the mortality for the spring months of 1888, however, we shall find that it is unusually high, and it is probable that this high death-rate has necessarily left all the fewer to die in the beginning of 1889, so that even in the presence of what may be looked on as unfavourable atmospheric conditions the death-rate is then considerably reduced.

Again, in connection with the low and high mortality in the winter months of 1887 and 1890 respectively, it will be found that in the former case the temperature is low, the mean and absolute range of temperature comparatively low, and the rainfall and number of rainy days very high, while in the latter case the temperature is comparatively high, and the relative humidity is also high, both of which would appear to be favourable in the winter months to a low death-rate. The mean and absolute range are, however, higher also, and the rainfall and number of wet days much less.

All these observations would lead to the conclusion that a season of low mortality in its relation to the state of the weather is coincident with a comparatively low barometer and temperature, low mean daily and absolute range of temperature, high rainfall, and large number of rainy days. The range of temperature is probably a more constant quantity than mere mean temperature. A high winter temperature may possibly be associated with low mortality, provided the range is low and number of wet days high; but if these do not accompany it, a high temperature may be attended by a higher death-rate than we find in connection with a low temperature if in the latter case the range be also low.

Remarks on Phthisis.

Bearing in mind that the above remarks apply to the mortality of a disease of a very chronic nature, and in no respect to its origin in relation to atmospheric conditions, it will be found :—

1. That low barometric pressure is generally associated with a low death-rate from phthisis.

2. That a comparatively low temperature in summer and autumn and a high temperature in winter are favourable to a low mortality, provided in all cases the mean and absolute range be low, and the rainfall and number of wet days be high. A low temperature during the winter months is, however, favourable to a low death-rate, if accompanied by a low range, high rainfall, and number of rainy days ; and conversely a high temperature in winter is not attended by a low mortality unless the range be also low, and the rainfall and number of wet days be high.

3. That the influence of the relative humidity is somewhat doubtful, although upon the whole a high humidity is connected with low mortality.

4. That the more uniform for the year the mean daily and absolute range of temperature, and amount of cloud, and the greater the humidity, rainfall, and number of rainy days, the lower will be the mortality.

5. That a year of heavy rainfall does not necessarily coincide with one of low death-rate, nor *vice versâ.* The latter depends rather on the regularity or otherwise with which the rain falls from one season to another: the more uniform the fall the less extreme will be the death-rate.

6. That a heavy rainfall in summer is favourable to a low mortality during the succeeding months of winter, provided it is followed by a fairly high fall in autumn and winter; but if after a heavy rainfall in summer the winter becomes very dry, the death-rate will rise unusually high. On the other hand, a high rainfall in winter is beneficial only when preceded by a high fall the previous summer.

7. That a period of large number of rainy days is especially connected with one of low mortality.

8. That a year of medium amount of cloud is associated with a fairly uniform mortality. The lower it is in winter the greater will be the death-rate, especially if preceded by a period of great amount of cloud in summer.

9. That the influence of the wind both in force and direction is somewhat doubtful.

PNEUMONIA.

Pneumonia—Queensland as a whole.—The mean monthly mortality curve of pneumonia for the whole of Queensland, for the quinquennial period 1887-91, shows a gradual rise for the first four months, a sudden rise for the next four, attaining its maximum in August, followed by an equally rapid fall the remaining four months of the year. The mortality is low in December, January, February, March, and April; medium in May, June, September, October, and November; and high in July and August.

On comparing this curve with the various meteorological readings, we find that the period of greatest mortality—winter—corresponds to the period of high barometer and high mean and absolute range of temperature, low temperature, rainfall, and number of rainy days, and low amount of cloud, the humidity and force of wind having apparently here no connection. It may be observed that while the month of July shows the lowest mean temperature for the year, the death-rate does not attain its maximum till August; that the latter month corresponds to lowest depression in the amount of cloud; and that while the mean and absolute range of temperature rise during the winter months along with the mortality, they continue still to rise for two or three months after the latter falls, and reach their maximum for the year in October.

During the first four months of the year there is a slight tendency for the mortality to increase each month in sympathy with a corresponding fall in temperature and amount of cloud, and rise in barometer, but inasmuch as the mean range of temperature remains stationary, and the absolute range falls a little, this increase in death-rate is small, and altogether the mortality is low for this period of the year. The first sudden rise is from April to May, and here we find that, in addition to a still further depression in temperature and amount of cloud, both the mean daily and absolute range of temperature for the first time rise considerably. Under similar conditions together with high barometer, low rainfall, low number of rainy days, and falling humidity, the mortality increases rapidly till July—the month of lowest mean temperature for the year. The latter now begins to rise, but inasmuch as its mean and absolute range go on to increase, the humidity to fall, and the amount of cloud to attain its minimum, the mortality rises still further to its maximum in August. The mean and absolute range now remain stationary for a couple of months, then along with the barometer begin to decline; the amount of cloud together with rainfall, humidity, and temperature meantime rises, and the death-rate falls, attaining its minimum for the year the following January.

MORTALITY PER 100,000 POPULATION PER ANNUM for the VARIOUS MONTHS of the YEAR.

Jan.	Feb.	March.	April.	May.	June.	July.	Aug.	Sept.	Oct.	Nov.	Dec.
32·4	34·8	38·4	39·6	57·6	63·6	94·8	99·6	66·0	60·0	63·6	40·8

66

The following will give the ratio of mortality per cent. for four periods of the year for pneumonia and phthisis :—

—	Jan. to Mar.	April to June.	July to Sept.	Oct. to Dec.	Total.
Pneumonia	15·2	23·2	37·7	23·9	100
Phthisis	23·9	23·9	28·7	23·5	100

showing in the case of pneumonia a much greater difference between the mortality of the summer and winter months than we find in connection with phthisis. In the former case it will be seen that for every 10 deaths that take place the first quarter there are 25 the third, whereas in the latter the ratio is 10 to 12, the mortality from phthisis, as has already been stated, varying but slightly from one month to another all the year round.

MORTALITY PER 100,000 POPULATION for the YEARS as a Whole.[1]

—	1887.	1888.	1889.	1890.	1891.
Pneumonia	50·4	70·3	60·8	65·9	43·4
Phthisis	130·8	140·1	128·6	135·7	127·3

RATIO PER CENT. for the YEARS as a Whole.

—	1887.	1888.	1889.	1890.	1891.	Total.
Pneumonia ...	17·2	23·8	21·3	23·0	14·7	100
Phthisis	18·7	21·4	20·1	21·0	18·8	100

It will here be observed that the rise and fall of pneumonia follows pretty much the same course as phthisis. The former, however, shows a somewhat greater variation from one year to another, and especially a greater fall in death-rate in 1891. In both cases the highest mortality takes place in 1888 and 1890, and the lowest in 1891. It may also be noted that for every 10 deaths from pneumonia there were on an average 23 from phthisis for the whole of Queensland during the last five years.

---

[1] The average mortality from pneumonia for the years 1887-91 is 54·1, and for the twelve years 1880-91 64·7 per 100,000 population per annum. In England the death-rate (1886-90) is 111·7, showing that for every 10 deaths from pneumonia in Queensland there are 20 in England according to population.

D

MEAN MONTHLY METEOROLOGICAL OBSERVATIONS for the YEARS as a Whole.

—	1887.	1888.	1889.	1890.	1891.
Barometer	30·016	30·074	30·055	29·986	30·031
Temperature	70·1°	71·7°	73·5°	71·7°	70·4°
Mean range	19·4°	21·6°	21·2°	19·9°	20·7°
Absolute range	35·4°	42·7°	42·0°	38·1°	37·8°
Humidity	71%	61%	63%	67%	63%
Rainfall	3·979	2·222	2·369	4·333	3·742
Rainy days	9	5	7	9	8
Cloud	3·9	2·9	3·7	4·2	4·1
Wind	11½	12½	12	12½	14½

In comparing these tables with each other it will be found that there is no constant connection between the meteorological observations with a *year as a whole* of high or low mortality. It is true that if the year 1887 be taken as representing a low death-rate—as it very fairly does—it is associated with a comparatively low barometer, low mean temperature, and low range of temperature, together with high relative humidity and rainfall, and large number of wet days; and that if 1888 be representative of a high mortality, as it most certainly is, the atmospheric conditions are in every respect the opposite of these. In 1890, however, the death-rate is almost equally high, and yet the barometric pressure is low, the humidity relatively high, and the rainfall and amount of cloud higher than in 1887. There remains, indeed, for that year only one element that seems to be connected with a high mortality—namely, a high absolute range of temperature.

Again, in 1891 it is impossible to associate the extremely low death-rate of that year with any peculiarity in the condition of the air, for except in low temperature, high amount of cloud, and high force of wind, the remaining elements follow generally a medium course.

In comparing the monthly or seasonal mortality, however, with meteorological observations, it is possible we may find a closer connection than seems apparent when the comparison is made from one year to another as a whole. Here we find that for the months of summer the death-rate is very high in 1889, and almost equally low in other years.

MORTALITY PER 100,000 POPULATION PER ANNUM in DECEMBER, JANUARY, and FEBRUARY.

1888-89.		Mean of Other Years.
56·8	...	29·7

The proportion of deaths in the summer of 1889 is thus nearly 2 to 1 that of other years. The following atmospheric conditions are coincident with the period of high and low mortality respectively:—

MEAN MONTHLY METEOROLOGICAL OBSERVATIONS—DECEMBER, JANUARY, and FEBRUARY.

	1888-89.		Mean of Other Years.
Barometer	29·965	...	29·874
Temperature	83·3°	...	79·4°
Mean range	23·7°	...	18·1°
Absolute range	44·2°	...	33·8°
Humidity	50%	...	68%
Rainfall	2·366	...	7·800
Rainy days	7	...	12
Cloud	3·7	...	5·4
Wind	11	...	12½

It is particularly evident from these figures that a summer of high mortality is distinctly associated with high barometric pressure, high temperature, and high range of temperature, mean and absolute, combined with low relative humidity, rainfall, number of wet days, and low amount of cloud; and that a low mortality is coincident with conditions exactly reversed to these.

As regards the winter months, it will be found the death-rate is high in 1888 and 1890, and comparatively low the remaining three years :—

MORTALITY PER 100,000 POPULATION PER ANNUM in JUNE, JULY, and AUGUST.

1888.	1890.	Mean of Other Years.
110·4 ...	123·6	65·7

there being, as will here be seen, about twice as many deaths in the two former years as on an average in the other three.

MEAN MONTHLY METEOROLOGICAL OBSERVATIONS, JUNE, JULY, and AUGUST.

—	1888.	1890.	Mean of Other Years.
Barometer...	30·192	30·112	30·129
Temperature	60·8°	60·9°	59·5°
Mean range	26·5°	21·9°	21·6°
Absolute range	45·6°	43·2°	40·4°
Humidity...	63%	71%	71%
Rainfall	·222	·727	1·961
Rainy days	2	5	5
Cloud	1·7	3·1	3·1

If the observations in connection with 1888 be compared with those in the third column, the one representing a winter of high mortality and the other one of low death-rate, it will be observed that in the former the barometer and mean and absolute range of temperature are high, while the relative humidity, rainfall, number of wet days, and amount of cloud are equally low. The influence of mean temperature upon the mortality in the winter months, on the other hand, appears somewhat doubtful, for although there is a low temperature and low death-rate in 1891 and a high temperature and high mortality in 1888 and 1890, the temperature for the same period in 1889 is the highest under observation, the death-rate being at the same time low. In connection with the high mortality in winter 1890, there does not appear from the observations above noted to be any reason why the rate should be specially high, or at least so high as in 1888, and yet it is somewhat higher. We may notice that the absolute range of temperature is comparatively high and that the rainfall is particularly low, but otherwise the atmospheric conditions seem to be quite consistent with a low mortality.

An explanation of this apparent anomaly is, however, probably obtained when reference is made to the meteorological observations from one month to another. We here find (1) that though the mean temperature of 1890 for the three months together—June, July, and August—be moderately high there is a more than usually sudden fall of 5·1 deg. from June to July, while the greatest fall of any other year (1888) is 3·4 deg., and in 1887 there is an actual rise; (2) that the mean and absolute range of temperature, though, as a whole, higher in 1888, take a more decided rise in 1890, and that possibly the effect of this rise is accentuated by the comparatively low range that obtains for several months together during the first half of the year.

It may be here noticed, not only that the rate in 1888 is high, but that it is more equally distributed over the various seasons, and shows what no other year does—three maximum curves of nearly equal height for June, August, and November. This is probably associated with a comparatively high mean daily and absolute range of temperature for the whole year, together with a steadily falling humidity and a greater number of months showing low rainfall, low number of wet days, and low amount of cloud than we find any other year.

Pneumonia—Brisbane District.—The mean monthly mortality curve of pneumonia for this district for the years 1887-91 is low for the months of January and February, rises rapidly from February to March, and thereafter more gradually to its maximum for the year in July, then falls pretty regularly till November, and rises somewhat in the month of December. The curve showing the admission rate to hospitals for acute lung diseases (pneumonia, bronchitis, and pleurisy) does not present the same rapid rise from February to March, nor the increase from November to December. This is probably accounted for by including bronchitis, which is seen, according to mortality statistics for the whole of Queensland, to be low both in March and December.

In connection with this rapid rise in pneumonia from February to March we may note that the mean temperature and amount of cloud begins slightly to fall; that in consequence of the latter and subsequent to periods of heavy rainfall evaporation goes on rapidly, thereby increasing the elastic force of vapour in the air, and the barometer takes a decided rise; that the absolute range of temperature also increases slightly; and that although the force of the wind does not appear to have a constant and direct influence on the death-rate it may be observed that it attains its maximum force during this month. The direction of the prevailing winds may also be here noted for the first three months of the year. Taking the mean for five years, the following shows the relation between east, south-east, and south winds during that period :—

E.	January,	observed	23	times,	forming	32	per cent.	of the whole.	
	February	,,	18	,,	,,	25			,,
	March	,,	17	,,	,,	24			,,
S.E.	January	,,	15	,,	,,	21			,,
	February	,,	16	,,	,,	25			,,
	March	,,	25	,,	,,	36			,,
S.	January	,,	9	,,	,,	13			,,
	February	,,	9	,,	,,	14			,,
	March	,,	18	,,	,,	26			,,

This shows that in January the prevailing wind is easterly. forming as it does 32 per cent. of the whole ; that in February the prevailing wind is equally divided between easterly and south-easterly, cach forming 25 per cent. of the whole; and that in March, while the easterly winds recede a little, the south-easterly winds largely predominate, forming 36 per cent. of the whole. The southerly winds also increase in frequency from 13 per cent. in January to 26 per cent. in March. At first sight this increase in south winds might lead us to partly attribute the increase in mortality in March to this particular cause, but further examination will show that this cannot be the case, for in March, 1888, the south winds prevail and yet the mortality for that period is unusually low.

Neither can it be said that a frequent change in the direction of the wind increases the mortality here, for although in February the wind is observed to change 63 times, and in March 70 times, we find it observed 71 times in January, when the mortality and admission rate are low. Perhaps, however, the high temperature in the latter month, together with other favourable conditions, may not have allowed this frequent change in the direction of the wind to affect the rate to the same extent as it might do later on in the presence of more adverse atmospheric conditions.

Under a rising barometer and mean and absolute range of temperature, and falling temperature, rainfall, number of rainy days, and cloud, and a tendency towards south-westerly and westerly winds, the mortality now continues high till May, then, with a sharp decrease in relative humidity together with an increase in the above conditions, rises to its maximum in July—the month in which the lowest temperature and lowest amount of cloud are recorded. It now begins to decline till November, rises somewhat in mortality—though not in admission rate—in December, and falls to its minimum for the year in January and February.

With regard to the death-rate between one season and another, we find that it corresponds very closely in the inverse ratio to the mean temperature, amount of cloud, rainfall, number of wet days, and force of wind. The following tables will show the mortality for the various months of the year according to population, and the ratio per cent. for four periods of the year:—

MORTALITY PER 100,000 POPULATION PER ANNUM for the VARIOUS MONTHS of the YEAR.

Jan.	Feb.	March.	April.	May.	June.	July.	Aug.	Sept.	Oct.	Nov.	Dec.
18·0	15·6	34·8	32·4	32·4	39·6	44·4	38·4	30·0	27·6	27·6	37·2

RATIO PER CENT. for FOUR PERIODS of the YEAR.

—	Jan. to Mar.	April to June.	July to Sept.	Oct. to Dec.	Total.
Brisbane	18·0	27·6	29·7	24·7	100
Queensland	15·2	23·2	37·7	23·9	100

It will be seen that the maximum rate takes place from July to September, and the minimum from January to March, although there is not so much difference in mortality between these two periods as for the whole of Queensland. In the latter case, as has been noted, there are 25 deaths in the third quarter to 10 in the first, whereas according to the above figures for the district of Brisbane the relative proportion is about 16 to 10 respectively.

The following table will give, for the various years as a whole, the usual total mortality from pneumonia as obtained from statistics supplied by the Registrar-General's Department, along with the admission rate to hospitals, for acute respiratory diseases :—

MORTALITY and ADMISSION RATE of PNEUMONIA and ACUTE RESPIRATORY DISEASES respectively PER 100,000 POPULATION for the YEARS as a Whole.

—	1887.	1888.	1889.	1890.	1891.
Admissions	121·2	94·8	100·8	126·0	121·2
Mortality	40·8	28·8	26·4	39·6	21·6

It will be observed here that in both mortality and admissions the rate is highest in 1887 and 1890,[1] and that while the latter is high also in 1891 the mortality is lower than that of any other year. This would tend to show that while acute lung diseases generally are very prevalent in 1891, they are, as shown by the low death-rate, of a mild type.. It will be found, however, on reference to the mortality curve, that this high admission rate occurs chiefly in October and November, and that possibly the higher temperature and other more or less favourable conditions of these months tend to bring about a greater proportion of recoveries than usual.

In comparing the mortality and admission rate with the meteorological observations for *the years as a whole* there does not appear to be any distinct and constant relation between them. It is probable that this will be more apparent in their monthly or quarterly relation to each other than when the comparison is based upon the mere mean observations for the extended period of twelve months.

As regards the monthly death-rate from one year to another, the principal feature in 1887 is the comparatively high maximum for

---

[1] In January, 1887, and March, 1890, Brisbane was visited by rather serious floods (*see* "Appendix"). The following figures, showing the admission and death rate during the winter following in each case, are somewhat suggestive :—

ADMISSION RATE for all RESPIRATORY DISEASES PER 100,000 POPULATION PER ANNUM.

AUGUST.		MAY, JUNE, JULY, AUGUST.	
1887.	Other Three Years.	1890.	Other Three Years.
441·6	262·8	435·9	276·0

MORTALITY from PNEUMONIA PER 100,000 POPULATION PER ANNUM for JULY, AUGUST, SEPTEMBER.

1887.	1890.	Other Three Years.
73·6	73·6	17·9

It will here be seen that the admission rate is almost double, and the death-rate fully four times, that of the mean of the other three years for these months. Moreover, the disease must have been of a malignant type, as the relative increase in mortality is double that of the admission rate.

three successive months, July, August, and September, and in connection with this it will be noticed that the admission rate for lung diseases generally is also particularly high in August. Leaving out of consideration, in the meantime, the equally high or higher mortality during the same months in 1890, the comparative magnitude of the rate in 1887 will be apparent when compared with that of the remaining three years. We here find that while in the former case the mean monthly mortality from pneumonia per 100,000 population is 73·6 per annum, in the latter it is only 17·9. From a meteorological point of view, we may note that the relative humidity takes a sudden drop from August to September, and so possibly keeping up the death-rate longer than is usual ; that the lowest mean temperature occurs not in July but in June, it being lower by 5 deg. than that of the same month of any other year; that in only one instance (1890) is it equally low in July; and that the mean temperature for the three months June, July, and August is under that of any other year for the same period. It is true that the mean temperature in July is higher than that in June; but the latter is absolutely the lowest under observation, while the former is still comparatively low.

As far as the direction of the wind is concerned in its relation to the high mortality for July, August, and September, it may be compared to that prevalent in 1891 where the death-rate is particularly low.

PREVALENCE of SOUTH, SOUTH-WEST, and WEST WINDS MONTHLY for JULY, AUGUST, and SEPTEMBER.

	SOUTH.		SOUTH-WEST.		WEST.	
	Times Observed.	Per cent.	Times Observed.	Per cent.	Times Observed.	Per cent.
1887	16	27	5	9	16	26
1891	17	26	22	34	8	13

It would appear from these figures that the more westerly the winds are the greater will be the death-rate in these months, but if the month of August alone be considered, when both the mortality and admission rate are high, we find that the west winds are observed only 5 times and form but 8 per cent. of the whole, whereas the south winds are more prevalent than during any other year. Moreover, during a similar period of high mortality in 1890, the south wind is observed on an average 16 times monthly, forming 29 per cent.; the south-west wind, 13 times, forming 23 per cent. ; and the west wind, 12 times, forming 21 per cent. of the whole. In the month of August, 1887, it may be mentioned that the east wind is observed 10 times, and it is quite possible that this frequent change from east to west and south-west may make persons more susceptible and less prepared to withstand the influence of a wind that is so changeable. This is supported somewhat by a consideration of the winds for the same month the following year, where the death-rate is also moderately high and where the east wind is observed 7 times during the month. The same, however, does not apply to 1890, and again in July, 1891, the east wind is observed 8 times, the mortality being at the same time low. In 1888 it will be noticed that there is a sharp rise in mortality in the month of June, and comparing this with the atmospheric conditions for that time, we find very high barometric pressure for this month and the two previous ones, high mean daily and absolute range of temperature, comparatively low humidity and rainfall, and low amount of cloud.

An examination of the winds in their direction may also be interesting, as compared to the mean of other years.

PREVALENCE of SOUTH, SOUTH-WEST, and WEST WINDS for the MONTH of JUNE.

—	SOUTH.		SOUTH-WEST.		WEST.	
	Times Observed.	Per cent.	Times Observed.	Per cent.	Times Observed.	Per cent.
1888	13	24	11	20	9	16
Other years ...	14	22	10	17	31	52

This would tend to show very plainly that the west winds, which are observed much less frequently than usual, do not at least contribute to the high mortality in the month of June, 1888. It would rather appear that westerly winds are favourable to a low death-rate, for in 1887 the death-rate for that month is below that of any other year, and here we find that the westerly winds are very prevalent, being observed 40 times, and forming 63 per cent. of the whole, while according to the above figures they are in June, 1888 (with a high mortality), observed only 9 times, and form but 16 per cent. of the whole. If, again, we take into consideration the prevalence of west winds in the previous month of May as having possibly some connection with the mortality in June, it will be found that they are observed, in 1888 and 1887, 6 and 18 times respectively, the death-rate being in the former case high and in the latter low. It may be stated at the same time that in 1888 the wind for the month of June is very wide in its range, for while the west, south-west, and south winds together form only 60 per cent. of the whole, the north, north-east, and south-east combined contribute 37 per cent. In the year 1887, on the other hand, where the mortality is low, the wind is very limited in its range, the west, south-west, and south together forming 97 per cent. of the whole for the month of June. It would appear, therefore, that a wide range in the direction of the wind during any particular month is more fatal in the case of pneumonia than where it blows steadily from a certain direction, even though that direction be from the west, south, or south-west. It is well to bear in mind, however, that the direction of the wind is but one element in the weather, and that the high death-rate in June, 1888, may be due far more to other unfavourable atmospheric conditions than to the influence of the wind alone.

In 1889 we have a whole year of a uniformly low mortality; and here it will be observed that the temperature of June, July, and August is not only nearly equal but higher than that of any other year; that the mean daily range of temperature is pretty low all through; that the relative humidity is comparatively high in the winter and early spring months; that the rainfall continues moderately high; and that the amount of cloud is more uniform and higher than that of any other year.

In 1890, the most important feature—indeed, for the whole period under observation—is the decided rise in mortality from pneumonia, as well as a simultaneous rise in the admission rate to hospitals from acute lung diseases for the month of July. For this month alone we find the death-rate is at the rate of 87·1, and for the three months July, August, and September, 73·6 per 100,000 population per annum, whereas for the other three years, 1888, 1889, and 1891 (and leaving out 1887, where the mortality is equally high) it is only 19·3 and 17·9 respectively. In looking for a possible cause for this in the state of the weather, we note that the temperature makes a rapid decline of 5·2 deg. from June to July, while the mean of 1888, 1889, and 1891 is only 1·2 deg., and in 1887 there is a rise of 1·2 deg.; that the temperature of the latter month (56·1 deg.) is the lowest on record, being lower by 2·2 deg. than the mean of the other four years combined; that the absolute range is high and shows a marked contrast to an unusually low range during the first five months of the year; that the rainfall is very low in May, June, and July, the average monthly fall for that time being ·978 inches, while the mean of the other four years combined is 2·916 inches; and that there is a greater fall in the amount of cloud from the first three months of the year to winter and spring than in any other year.

The influence of the direction of the wind here is somewhat doubtful. The south is the prevailing wind, it being observed 20 times and forming 35 per cent. of the whole, but in the month of July, 1889, it is equally prevalent and yet the mortality is low. The west wind, again, occurs 15 times and forms 25 per cent. of the whole, and in 1889 and 1891, where the death-rate is low, it is on an average observed 6 times and forms only 9 per cent. of the whole. It would thus appear as if the high mortality had some connection with the west winds, but in 1888 they are equally common, and in 1887 much more so, forming as they do in the latter case 46 per cent. of the whole, and yet in both these instances the death-rate is much below that of July, 1890. Should the winds prevailing in June be taken into account as bearing possibly on the mortality for July, the west and south-west winds are even less common as compared to other years, while the south is the prevailing wind, forming 36 per cent. of the whole, the mean for the other four years being only 19 per cent.

In 1891 we find that the mortality in the months of March, April, and May is not only unusually high, but higher than during any of the winter months, the death-rate in July being lower than that of any other month of the year. In connection with this early and abnormal rise we note :—That the barometer ascends very rapidly; that although

the temperature in March is unusually high, it falls very rapidly in
April and May; that the absolute range of temperature begins to rise
in January and continues to rise up till May; that the relative
humidity is low during the first four months of the year; and that the
usual amount of rain does not fall in the summer months, the
average monthly fall for January, February, and March being 3·367
inches, against 9·016 for the remaining four years. As the mortality
for these months is high also in 1887 and low in 1888, a comparison
between these as regards the direction of the wind will be interesting.

—	SOUTH WIND. Times Observed.			SOUTH-WEST AND WEST COMBINED. Times Observed.		
	March.	April.	May.	March.	April.	May.
1887	14	50	32	0	2	22
1891	18	30	37	4	10	11
1888	23	19	18	9	0	10

This would appear to show that in March neither of these winds
are objectionable; that in April as well as in May the south winds are
unfavourable, while the influence of the west and south-west winds is
doubtful. During these three months also the wind is observed to
change monthly in 1887, 65 times; in 1888, 59 times; and in 1891, 74
times; showing probably that a frequent change in the direction of
the wind coincides with increase in the mortality.

With regard to the abnormal fall in death-rate for the month of
July, it will be observed that as compared to the month of May the
barometric pressure is low; that the mean range of temperature is
not much higher than that observed in the early months of the
year; that the absolute range is low; and that the rainfall, though low
in summer, has continued moderate during the months of winter. In
forming an estimate of the influence of the winds, a comparison may be
made with the same month of the previous year, where the mortality is
very high. .

—	SOUTH. Times Observed.	SOUTH-WEST. Times Observed.	WEST. Times Observed.
1890	20	15	15
1891	18	30	5

According to these figures, the absence of west and prevalence of
south-west winds are favourable in the month of July to a low
mortality from pneumonia.

BRONCHITIS.

THE mean monthly curve of mortality from bronchitis for the whole of Queensland for the quinquennial period 1887-91 rises gradually from a low minimum in January to its maximum for the year in June and July, and then falls steadily during the remaining months of the year.

The following will show the death-rate from bronchitis, and for comparison that also from pneumonia, for each month of the year according to population :—

MORTALITY PER 100,000 POPULATION PER ANNUM for EACH MONTH of the YEAR.

—	Jan.	Feb.	Mar.	April.	May.	June.	July.	Aug.	Sept.	Oct.	Nov.	Dec.
Bronchitis ...	27·6	31·2	40·8	48·0	39·6	66·0	70·8	61·2	55·2	37·2	38·4	32·4
Pneumonia ...	32·4	34·8	38·4	39·6	57·6	63·6	94·8	99·6	66·0	60·0	63·6	40·8

There is here a very close connection between the rise and fall of bronchitis and pneumonia, with this exception, that while the former attains its maximum in July the latter rises still further in August.

RATIO PER CENT. for FOUR PERIODS of the YEAR for BRONCHITIS, PNEUMONIA, and PHTHISIS.

—	Jan. to Mar.	April to June.	July to Sept.	Oct. to Dec.	Total.
Bronchitis	18·2	28·1	34·0	19·7	100
Pneumonia	15·2	23·2	37·7	23·9	100
Phthisis	23·9	23·9	28·7	23·5	100

It will be observed that the proportion of deaths from bronchitis is relatively greater in the first two quarters, and equally less in the last two quarters of the year, than we find in connection with pneumonia. In the case of bronchitis there are 18 deaths from July to September for every 10 from January to March, whereas in pneumonia the ratio is 25 to 10 respectively.

MORTALITY PER 100,000 POPULATION for the YEARS as a Whole.[1]

—	1887.	1888.	1889.	1890.	1891.
Bronchitis	44·2	48·9	42·4	49·8	44·2
Pneumonia	50·4	70·3	60·8	65·9	43·4
Phthisis	130·8	140·1	128·6	135·7	127·3

[1] The average annual mortality from bronchitis for the years 1887-91 is 45·9, and for 1880-91 44·7, per 100,000 population. The death-rate in England (1886-90) is 210·9, or nearly five times greater than that of Queensland in proportion to population.

RATIO PER CENT. for the YEARS as a Whole.

—	1887.	1888.	1889.	1890.	1891.	Total.
Bronchitis	19·2	21·3	18·5	21·8	19·2	100
Pneumonia	17·3	24·2	20·9	22·7	14·9	100
Phthisis	19·8	21·1	19·4	20·5	19·2	100

These figures will show that the years of highest mortality from bronchitis, pneumonia, and phthisis are the same—namely, 1888 and 1890; that although there is a marked decline in pneumonia in 1891, the number of deaths from bronchitis and phthisis is nearly equal to the average of other years; and that generally the relative proportion of deaths from bronchitis, from one year to another, resembles very closely that from pneumonia, but more especially that from phthisis.

If again we take the months generally of highest mortality—June, July, and August—and compare the death-rate from bronchitis, pneumonia, and phthisis, during these months, from one year to another, the following will show the relation they bear towards one another :—

MORTALITY PER 100,000 POPULATION PER ANNUM for JUNE, JULY, and AUGUST.

—	1887.	1888.	1889.	1890.	1891.
Bronchitis	46·8	81·6	55·2	82·8	64·8
Pneumonia	70·8	110·4	74·4	123·6	52·8
Phthisis	123·6	163·2	128·1	164·4	130·8

RATIO PER CENT. for JUNE, JULY, and AUGUST.

—	1887.	1888.	1889.	1890.	1891.	Total.
Bronchitis	14·1	24·6	16·7	25·0	19·6	100
Pneumonia	16·4	25·5	17·2	28·7	12·2	100
Phthisis	17·4	23·0	18·0	23·2	18·4	100

These very fairly show that a winter of high or low mortality from bronchitis is almost equally high or low for pneumonia and phthisis, the only exception being in 1891, where the death-rate from pneumonia is low while that from bronchitis and phthisis is moderately high. Why this exception should exist it is not easy to say. We may observe that the low temperature in the winter months of 1891 is associated with a low absolute range of temperature and moderate rainfall, and as it has been noticed that pneumonia is almost necessarily associated with *high* absolute range of temperature under similar atmospheric conditions, and that in one or two districts (Rockhampton, Boulia, and Blackall) these conditions are more especially associated with an increased number of cases of bronchitis, it would appear that we have here an explanation of the high mortality for one, and low death-rate for the other. It will be found, however, that similar conditions also obtain in the winter months of 1887, and yet the mortality is comparatively low, not only from pneumonia, but also from bronchitis. It may therefore be fairly concluded, that as the death-rate from bronchitis and pneumonia rise and fall together, not only from one year to another as a whole, but during the winter months of each year, they are associated with pretty similar atmospheric conditions.

REMARKS ON PNEUMONIA AND BRONCHITIS.[1]

1. That a low mortality generally occurs in the summer months, and is connected with low barometric pressure and low mean daily and absolute range of temperature, high temperature, rainfall, number of rainy days, and high amount of cloud, together with, for the eastern portion of the colony, east and south-east winds.

2. That it rises in the autumn months in accordance with a rising barometer and mean daily and absolute range of temperature, a falling temperature, rainfall, number of wet days, and cloud, and to a tendency to south-east and south winds.

3. That it rises to its maximum in the winter months on account of a continued high condition of the barometer, a fall in relative humidity of the air, a still further depression in temperature, rainfall, and amount of cloud, a further elevation in mean daily and absolute range of temperature to their maximum for the year, and, for the eastern portion of the colony, south, south-west, and west winds.

4. That it falls in the months of spring in sympathy with a reversal of all those condition of the atmosphere above stated (except that of relative humidity, which falls to its minimum for the year), and with a return to south-east, east, and south winds.

5. That the sooner the absolute range of temperature rises, and the more rapidly the temperature falls in autumn, the greater will be the mortality for that period, and if combined with a rapid rise in barometric pressure it may even then attain its maximum for the year.

6. That a high temperature in the summer months is associated with a high death-rate if attended by low rainfall and low relative humidity, and high mean and absolute range of temperature. That a low temperature in winter is not associated with a high mortality provided its range is low, the rainfall moderate (Brisbane, 1887, excepted), and the relative humidity comparatively high; a rapid fall of several degrees from one month to another in winter is more intimately connected with a high death-rate than mere mean low temperature for several months together. The mortality attains a high maximum in winter, if a rapid decline in temperature is accompanied by low rainfall and sudden rise in absolute range.

7. That the death-rate will be all the higher in winter, the greater the difference in absolute range of temperature, rainfall, and amount of cloud between summer and winter.

---

[1] These remarks apply mainly to pneumonia.

8. That a low rainfall in summer and autumn does not seem to affect the mortality so much as a low rainfall in winter, possibly because of the other atmospheric conditions being more favourable, and because the fall is never so low as is generally observed during the winter months.

9. That the influence of the force and direction of the winds cannot certainly be ascertained by mean monthly observations. In some instances for the Brisbane district where the death-rate is high, the south is the prevailing wind, and in others the west. The latter is probably more frequently associated with a high mortality.

10. That the death-rate will be all the higher the more frequently the wind changes, and the greater its range, during any particular period.

11. That in a general manner pneumonia is associated with atmospheric conditions that allow of greater loss of heat by evaporation and radiation, than by conduction.

12. And that, in order to counteract as much as possible the effect of these atmospheric conditions, our clothing during the autumn and winter months should be of such material as to retard the loss of heat by evaporation, more than by conduction.

RESPIRATORY DISEASES.

Acute Respiratory Diseases—Rockhampton Division.—The mean monthly mortality curve of acute respiratory diseases (pneumonia, bronchitis, and pleurisy) for this district for the quinquennial period 1887-1891 shows a maximum rise for the months of June, July, and August, and a fairly uniform and low rate for the remaining nine months of the year.

MEAN MORTALITY PER 100,000 POPULATION PER ANNUM for the VARIOUS MONTHS of the YEAR.

Jan.	Feb.	March.	April.	May.	June.	July.	Aug.	Sept.	Oct.	Nov.	Dec.
157·2	157·2	100·8	135·6	171·6	328·8	444·0	418·8	200·4	171·6	236·4	178·8

RATIO PER CENT. for FOUR PERIODS of the YEAR.

January to March.	April to June.	July to September.	October to December.	Total.
15·4	23·6	39·2	21·8	100

The proportion of deaths, according to the above, from one period of the year to another, is pretty much the same as in the case of pneumonia for Queensland as a whole, there being, on an average of five years, 25 deaths from July to September to every 10 from January to March.

In comparing the mortality from one year to another, the following tables will give the rate according to population, and the meteorological observations connected therewith :—

MORTALITY per 100,000 POPULATION for the YEARS as a Whole.

1887.	1888.	1889.	1890.	1891.
180·0	356·4	225·6	226·8	145·2

showing that the years of highest and lowest death-rate are 1888 and 1891 respectively.

MEAN MONTHLY METEOROLOGICAL OBSERVATIONS for the YEARS as a Whole.

—				1887.	1888.	1889.	1890.	1891.
Barometer	...	...	...	...	30·094	30·080	30·028	30·074
Temperature	...	...	...	71·2°	71·3°	72·4°	72·1°	71·6°
Mean range	...	...	...	13·9°	14·7°	13·9°	17·7°	18·9°
Absolute range	...	...	...	28·4°	29·4°	29·0°	34·4°	34·1°
Humidity	...	...	...	75%	74%	74%	71%	67%
Rainfall	...	...	...	3·833	3·883	2·986	6·443	4·256
Rainy days	...	...	...	10	6	10	13	10
Cloud	...	...	...	4·1	3·8	4·7	5·1	4·8
Wind	...	...	...	11½	13	12	12½	12

It will here be observed that the meteorological readings are fairly uniform from one year to another, and do not apparently present any particular feature indicative of either a high or low mortality. There is at least nothing in 1888 except high barometric pressure, low number of wet days, and low amount of cloud, to account for the high mortality of that year. It may be noticed also that the mean and absolute range of temperature are particularly high and the relative humidity is low during the year of lowest death-rate, 1891. It is probable that a more satisfactory interpretation may be obtained when the mortality is examined monthly, or quarterly, from one year to another.

MORTALITY PER 100,000 POPULATION PER ANNUM for the MONTHS of JUNE, JULY, and AUGUST.

1887.	1888.	1889.	1890.	1891.
223·2	637·2	312·0	540·0	241·2

These figures show that for the months of June, July, and August the death-rate is low in 1887 and 1891, and high in 1888 and 1890. The following table will give the corresponding meteorological observations :—

MEAN MONTHLY METEOROLOGICAL OBSERVATIONS for JUNE, JULY, and AUGUST.

—	1887.	1888.	1889.	1890.	1891.
Barometer	...	30·204	30·151	30·146	30·152
Temperature	61·3°	61·5°	61·8°	61·6°	61·0°
Mean range	16·9°	18·9°	15·6°	20·8°	23·3°
Absolute range	33·4°	36·1°	31·1°	42·3°	38·8°
Humidity	78%	78%	81%	78%	72%
Rainfall	2·506	·152	1·900	2·247	2·902
Rainy days	7	3	6	6	5
Cloud	3·4	2·1	3·4	3·6	2·7

We again find here that the year of highest mortality in the winter months is particularly associated with very high barometric pressure, low rainfall, and number of wet days, and low amount of cloud, while the mean and absolute range, and also the mean temperature and humidity, follow a medium course.

In 1887 the mortality is, from one month to another, uniformly low, and we find although the temperature is low in June that the mean daily and absolute range are also low, that the rainfall continues moderate all the year, and that the amount of cloud is high.

In 1888 the death-rate does not rise to the same maximum as observed in July and August, 1890, but we find here what is not observed any other year, an equally high rate for the months of June, July, August, and November, thus giving it as a year the premier position, in point of mortality, for acute lung diseases.

From a meteorological point of view it will be observed that the barometer rises suddenly from February till April, remains high till September, and then falls slowly; that the temperature is low in the summer months, and falls steadily to a lower level in July than any other year except 1890; that after a very heavy rainfall of 32 inches in February there is practically none during the remaining ten months of the year, the average per month being only ·685 inch, whereas the mean for the other four years for the same period is 11·518 inches; that the average number of rainy days per month during the same time is four for 1888, and nine for the other four years; and that the amount of cloud is lower than that of any other year. As regards the mean daily and absolute range of temperature and relative humidity, these do not show any peculiarity indicative either of a high or low mortality.

In 1889 the mortality never attains a high maximum, and it will be noticed that the temperature in summer is high, and though low in June only falls half-a-degree from June to July; that the mean daily and absolute range are low; that the humidity and amount of cloud are high; and that the rainfall, though low in summer, continues moderate during the winter months.

In 1890 the mortality rises very high in July and August, and it will be observed that there is an abnormal fall of temperature from June to July of 5 deg.—the average for 1888, 1889, and 1891 being 1·7 deg., and in 1887 there is a rise of ·8 deg.; that the temperature in July is below that for the same month of any other year, being 58·6 deg., while the mean for the other four is 60 deg., and for 1889 where the mortality is low, 61·2 deg.; that the absolute range of temperature rises suddenly from May to June, and during the three winter months is higher than that of any other year, being on an average 42·3 deg., while that of the other four years is 34·8; that the rainfall is heavy in January, February, and March, the average number of inches per month being 17, while the mean of the other four years is 8, the number of rainy days being for the same time 24, while for the other years it is only 13; that after March the rainfall is moderate, and number of rainy days slightly above the average; and that the west winds are more prevalent than in any other year.

It may here be noticed that the high mortality in 1888 is connected with somewhat different conditions to those observed in 1890. In the former year it is associated with a year of exceedingly low rainfall (subsequent to a heavy fall in March), and with conditions dependent thereon, such as high barometric pressure, and low amount of cloud. In the latter year, on the other hand, the rainfall is very high for the three consecutive months January, February, and March, and continues comparatively high during the remainder of the year, the barometric pressure being at the same time low and amount of cloud high. In this case, however, we find that the temperature in July is low, and after several months of low absolute range the latter now suddenly rises beyond that observed any other year.

In 1888 the mortality is high because of the low rainfall, low amount of cloud, and high barometric pressure, the temperature

E

being at the same time moderately low. In 1890 the death-rate is high, not because the rainfall is high, but because this heavy rainfall is accompanied by an exceptionally low temperature in July, and very sudden and excessive rise in range of temperature. In the one case, loss of heat possibly takes place by radiation and evaporation, in the other by conduction.

In 1891 the death-rate is low, and here it may be noted that the temperature only falls 1·5 deg. from June to July, and is accompanied by a fair amount of rainfall during the winter months.

Acute Respiratory Diseases—Townsville.—Here the returns unfortunately are incomplete, but it may be observed at least that the year of highest mortality, 1888, corresponds to one of high barometric pressure, low rainfall, and number of wet days, and low amount of cloud.

Respiratory Diseases—Cooktown Division.—The mean monthly curve of admissions to hospitals for the years 1887-91 is low in January and February, high in July, and medium the remaining months of the year. The following will give the mortality and admission rate to hospitals from one year to another as a whole according to population.

MORTALITY and ADMISSION RATE PER 100,000 POPULATION for the YEARS as a Whole.

—	1887.	1888.	1889.	1890.	1891.
Admissions	194·4	480·0	286·8	225·6	297·6
Mortality	424·8	463·2	542·4	471·6	435·6

It will be observed that 1888 is the year of highest number of admissions, and 1889 the year of highest death-rate. It is possible that in the admissions of the former year a number of sub-acute and chronic cases are included, and as their illness is of some duration, they go to increase the death-rate of the year following.

In comparing these with the meteorological readings for the years as a whole we observe a high absolute range of temperature and low amount of cloud in 1888 ; otherwise, the conditions are more or less of a medium character.

As regards the monthly curve from one year to another, the only thing that particularly requires attention is a high number of admissions in March, 1888, and also (although this curve is not shown in the chart) a high mortality in August of the same year. In the former case the number of admissions from respiratory diseases to the various hospitals in this district is at the rate of 1,213·2 per 100,000 population per annum, while that of other years is on an average only 193·1, or in the proportion of fully 6 to 1. In the latter case the mortality for the month of August is at the rate of 1,323·6 per 100,000 population, while in other years it is 551·9, or in the proportion of 2·4 to 1. In order to include both these months, the mean monthly meteorological observations from March to August may be taken and compared to those of other years for the same period :—

MEAN MONTHLY METEOROLOGICAL OBSERVATIONS from MARCH to AUGUST.

	1889.		Mean of Other Years.
Temperature	75·8°	...	76·4°
Mean range	11·4°	...	10·4°
Absolute range	25·6°	...	22·0°
Humidity	73%	...	73%
Rainfall	3·363	...	4·874
Rainy days	10	...	11
Cloud	3·2	...	4·4

This will show fairly that during the period of high mortality and admission rate the mean and absolute range of temperature are comparatively high, while the mean temperature, rainfall, and amount of cloud are equally low. Moreover, in making a more detailed examination it will be noted that in July the temperature is especially low, being 70·3 deg., while the mean for other years is 73·3 deg.; that the mean range is above the average, being 13·8 deg., while that of other years is only 10·6 deg.; that the absolute range in June, July, and August is on an average 28 deg., while for other years it is 24·2 deg.; that in July and August the mean monthly rainfall is only ·310 inches, while that of other years is 1·400 inches; and that during these months the amount of cloud is only 2·6, and that of other years 4.

Respiratory Diseases—Normanton Division.—The population for this district is so small (3,434 in 1888 and 4,962 in 1891) that it is impossible to form an accurate idea of the mortality or admission rate in their relation to atmospheric conditions. The following is as near as possible the yearly death-rate for all respiratory diseases according to population.

MORTALITY PER 100,000 POPULATION for the YEARS as a Whole.

1888.	1889.	1890.	1891.
840·0	456·0	80·4	121·2

There is here a very marked difference in mortality from one year to another, the proportion in 1888 to that in 1890 being fully 10 to 1. Unfortunately also the meteorological observations are not complete, but at least it may be said that in the year of highest mortality no rain whatever falls from April to November, that the amount of cloud during that time is exceedingly low, and the barometric pressure equally high.

Respiratory Diseases—Cloncurry and Hughenden Division.— The population for this large district being also very small (4,408 in 1888 and 5,434 in 1891), one can hardly form even an approximate estimate of the connection between the mortality and the state of the weather. The yearly death-rate from respiratory diseases according to population and the atmospheric conditions connected therewith are as follow:—

MORTALITY PER 100,000 POPULATION for the YEARS as a Whole.

1888.	1889.	1890.	1891.
294·8	228·0	212·4	242·4

MEAN MONTHLY METEOROLOGICAL OBSERVATIONS for the YEARS as a Whole.

—	1888.	1889.	1890.	1891.
Temperature	76·5°	78·7°	75·5°	72·8°
Mean range	27·5°	26·2°	24·9°	24·9°
Absolute range	44·9°	48·6°	46·0°	42·7°
Humidity	49%	53%	59%	54%
Rainfall	1·861	1·336	2·168	3·474
Rainy days	3	4	6	5
Cloud	1·5	2·2	2·9	2·8
Wind	13	12	12½	12½

It will be noticed that the year of highest mortality is 1888, and that this corresponds with high mean range of temperature, low humidity, rainfall, and number of wet days, and low amount of cloud. The mean temperature and its absolute range follow a medium course.

In following the monthly curve it will be seen that the higher death-rate in 1888 is not caused by any unusual rise in any particular month or months, but rather by a comparatively and slightly higher rate all the year. It will really be found that the highest mortality during the four years under review occurs in July and August, 1891, although for the year as a whole it is lower than in 1888. The meteorological readings associated with the high death-rate for these two months are a low temperature (the mean for the three months of June, July, and August being 59·1 deg., while that of other years is 64·1 deg.), low relative humidity, and low amount of cloud. It may also be noted that the rainfall, though low, is relatively *higher* than that of any other year.

Respiratory Diseases—Boulia and Blackall Division.—The mean monthly curve of admission to hospitals for respiratory diseases for the years 1888-1891 shows a maximum rate in the months of July and August and a fairly uniform rate the remaining months of the year.

RATIO PER CENT. for FOUR PERIODS of the YEAR.

—	Jan. to Mar.	April to June.	July to Sept.	Oct. to Dec.	Total.
Admissions	22·1	20	33·5	24·4	100
Mortality	16·2	27	33·3	23·5	100

This will show that in the first quarter the number of admissions is high as compared to the mortality, that in the second this is reversed, and in the third and fourth the ratio is about the same in both cases.

MORTALITY and ADMISSION RATE PER 100,000 POPULATION for the YEARS as a Whole.

—	1888.	1889.	1890.	1891.
Admissions	538·8	511·6	874·8	529·2
Mortality	175·2	192·0	240·0	171·6

According to these figures the year of highest death-rate and number of admissions to hospitals is 1890, and the following will, for comparison, give the meteorological observations for the years as a whole:—

MEAN MONTHLY METEOROLOGICAL OBSERVATIONS for the YEARS as a Whole.

—	1888.	1889.	1890.	1891.
Temperature	74·5°	75·6°	70·8°	70·3°
Mean range	31·2°	29·5°	23·4°	25·1°
Absolute range	52·5°	66·2°	43·7°	43·9°
Humidity	47°/₀	52°/₀	57°/₀	60°/₀
Rainfall	1·341	1·151	2·599	2·503
Rainy days	3	4	5	5
Cloud	1·8	2·5	3·3	3·2
Wind	18	13½	13½	15½

It will be noticed that during the year of highest mortality and admission rate the temperature is low, and that with this exception all the other atmospheric conditions are the reverse of what is usually observed in connection with a high death-rate from respiratory diseases, for we find the range of temperature low, and the relative humidity, rainfall, number of wet days, and amount of cloud comparatively high.

In referring to the monthly curve of that year it will be found that the death-rate is high for the three consecutive months August, September, and October, and the admission rate for July, August, and September. Here again the atmospheric conditions are exceptional in their character. If we take the months of June, July, and August, it will be found that the barometer is low, the average being 30·110, while that of the other three years is 30·153, that the mean range of temperature is low, 23·7 deg., while that of other years is 29·4 deg.; that the absolute range is also low, 48·6 deg., while that of other years is 52·2 deg.; that the humidity is relatively high, 67 per cent., while that of other years is 63 per cent.; that the rainfall, although about the same as the average of other years for these three months, shows a much higher general fall from the beginning of the year; that the same applies to the number of wet days; and that the amount of cloud is high, 3·0, while that of other years is 1·8. As regards the mean temperature, it will be noted that it falls 4·5 deg. from June to July, the mean fall for other years being 2·2 deg., and that in July it is lower than that of any other year, 53·7 deg., the average for other years being 55·8. With the exception therefore, of temperature, the remaining atmospheric conditions generally throughout Queensland accompany a period of low mortality, while in this instance they are associated with a high mortality from respiratory diseases.

It may be mentioned that, according to the hospital returns, the high admission rate is made up chiefly by a decided increase in bronchitis, catarrh, and influenza, and it would therefore seem probable that, while pneumonia is mainly associated with low temperature and low rainfall, the more catarrhal forms of acute lung complaints are prevalent rather under a relatively cold and humid atmosphere. It is well to bear in mind that for this large inland division a high rainfall seems to be abnormal, and it is possible that a heavier fall than usual may raise at least the admission rate to hospitals in districts where outdoor life is largely enjoyed, and where one is prepared in matters of clothing for conditions that favour the loss of heat by radiation but not those by conduction.

Respiratory Diseases—Roma and Thargomindah Division.—The mean monthly curve of mortality for this division for the years 1887-1891 is low in January and February, high in July, and fairly low the remaining months of the year.

RATIO PER CENT. for FOUR PERIODS of the YEAR.

January to March.	April to June.	July to September.	October to Dec.	Total.
17·4	28·7	31·8	22·1	100

These figures show that there are nearly twice as many deaths from July to September as from January to March. The following will give the general death-rate and admission rate to hospitals for the years as a whole, according to population.

MORTALITY and ADMISSION RATE PER 100,000 POPULATION for the YEARS as a Whole.

—	1887.	1888.	1889.	1890.	1891.
Admissions	308·4	220·8	333·6	332·4	298·8
Mortality	199·2	283·2	178·8	241·2	210·0

It will be noticed that the year of lowest number of admissions to hospitals—1888—is also the one of highest mortality for the district. It follows, if the same proportion of cases that occur in general practice are always sent to hospital, either that a great number of the cases admitted the year previous are of a chronic nature, and so increase the death-rate, not of 1887, but 1888; or that those admitted the latter year, though fewer in number, are more malignant in their type, and a greater proportion of them die. The following are the meteorological observations for the years as a whole :—

MEAN MONTHLY METEOROLOGICAL OBSERVATIONS for the YEARS as a Whole.

—	1887.	1888.	1889.	1890.	1891.
Barometer	...	30·189	30·076	30·019	30·076
Temperature	66·7°	67·9°	70·5°	68·1°	67·7°
Mean range	24·8°	28·2°	24·1°	22·4°	21·8°
Absolute range	46·9°	51·4°	49·4°	44·2°	44·6°
Humidity	67%	50%	61%	66%	63%
Rainfall	2·244	1·031	1·814	3·503	2·717
Rainy days	5	3	5	6	5
Cloud	2·9	2·4	4·0	3·9	3·8
Wind	9½	11½	12½	16	17

This will show that the year of highest mortality (which after all, in a district where hospitals are far apart, is a more certain guide than the admission rate) corresponds to high barometric pressure and high mean daily and absolute range of temperature, combined with low relative humidity, rainfall, and number of wet days, and low amount of cloud. The temperature also is comparatively low, though not so low as in 1887.

In examining the monthly curve it will be found that the high mortality of 1888 occurs in July, August, and September. From a meteorological point of view it will be found that the barometer is high, the mean for June, July, and August being 30·230, against 30·173 for other years; that the mean daily and absolute range of temperature are high, being 31·6 deg. and 50·7 deg., while the average of other years is 23·7 deg. and 43·9 deg. respectively; that the relative humidity is low, 57 per cent., and in other years 75 per cent.; that the rainfall is low, ·066 inch, and in other years 1·416 inches; that the number of wet days is very low also, 1, and the average monthly number in other years, 4; that the amount of cloud is low, 1·8, and in other years 3·7 ; and that the velocity of the wind is also somewhat low, being at the rate of 11 miles, while the average of other years is 12½ miles per hour.

It should also be noted that during these months the admission rate is low or at least does not rise in the same proportion as the death-rate, showing possibly, as previously mentioned, that the diseases are of a more than usually *malignant* type. Whether this is due in any way to the existing atmospheric conditions of the time it is impossible to say, but it may be observed in connection with this that in June, July, and August, 1890, and in March, April, and May, 1891, the mortality is low relatively to the admission rate — that is to say, the cases appear to be of a *milder* type—and the atmospheric conditions are in some respects different, the barometer and mean and absolute range of temperature being comparatively low, while the relative humidity, rainfall, and amount of cloud are equally high.

Respiratory Diseases—Darling Downs.—The mean monthly mortality curve of respiratory diseases for this district for the quinquennial period 1887-91 is low from October to February, high in May and June, and of a medium height the remaining months of the year.

RATIO PER CENT. for FOUR PERIODS of the YEAR.

—	Jan. to Mar.	April to June.	July to Sept.	Oct. to Dec.	Total.
Admissions	25·6	26·8	25·1	22·5	100
Mortality...	19·7	37·3	27·1	15·9	100

It will be observed that both the greatest number of admissions to hospital and greatest mortality for the whole district take place in the second quarter of the year. In the former case, however, the ratio per cent. is pretty nearly the same from one quarter to the other, whereas in the latter the proportion of deaths is very much greater from April to June than during any other similar period of the year.

The following tables will show the mortality and admission rate according to population for the various years as a whole, together with the meteorological observations connected therewith :—

MORTALITY and HOSPITAL ADMISSIONS PER 100,000 POPULATION for the YEARS as a Whole.

—	1887.	1888.	1889.	1890.	1891.
Admissions	380·4	511·2	321·2	478·8	247·2
Mortality...	187·2	135·6	156·0	133·2	164·4

MEAN MONTHLY METEOROLOGICAL OBSERVATIONS for the YEARS as a Whole.

—	1887.	1888.	1889.	1890.	1891.
Temperature	61·5°	62·8°	63·4°	62·8°	62·0°
Mean range	19·3°	24·8°	21·5°	21·7°	21·7°
Absolute range	36·5°	46·7°	44·9°	42·3°	41·5°
Humidity	79°/₀	70°/₀	73°/₀	74°/₀	74°/₀
Rainfall	4·005	1·723	3·052	3·743	3·329
Rainy days	10	6	10	12	10
Cloud	3·2	3·5	4·5	5·1	4·5
Wind	8	10½	12	12	13½

We here find that, while the highest number of admissions is recorded in 1888, the death-rate is comparatively very low, and that in this respect it is the reverse of what is observed for the division which Roma and Thargomindah represent. If we compare these with the atmospheric conditions for that year it will be noticed that the mean and absolute range of temperature are high, and that the relative humidity, rainfall, number of wet days, and amount of cloud are comparatively low.

It would appear as if these conditions were favourable to the development of respiratory diseases, but that being of a mild type a comparatively large proportion of the cases recover, and if that be so, it is contrary to what is noticed for the Maranoa district. Moreover, in 1889 and 1891 the admission rate is low and the mortality high, and here again we find a low range of temperature, high humidity and high rainfall, and thus so far supporting the above statement. It should be mentioned, however, that in 1890 the admission rate and mortality are high and low respectively as in 1888, and yet the atmospheric conditions are pretty similar to those of 1891.

If we refer to the monthly curve from one year to another, it will be found in 1887 that the admission rate is particularly high in April, and the mortality comparatively high from March to June. Unfortunately the meteorological observations are not complete, but at least it may be noted that the temperature falls suddenly from March till June, and that the mean daily and absolute range, though still low, take an equally sudden rise.

In 1890 the mortality is fairly high, and the admission rate for all respiratory diseases specially high in July and August, and here it will be found that the temperature takes a decided fall of 6·6 deg. from June to July, while in 1887 there is a rise of 1·7 deg., and in other years a fall of only ·8 deg.; that the temperature of July is below that of any other year—47·6 deg.—while the mean for other years is 49·6 deg.; that the mean and absolute range of temperature, though not so high as in 1888, are pretty nearly so, and take a very sharp and decided rise from a low range during the early part of the year; that the high rainfall in summer is succeeded by a low rainfall in winter; and that the amount of cloud and force of the wind, though not so low as in 1888, are still comparatively low. In some respects, therefore, the atmospheric conditions bear a fair resemblance to those of 1888 during the winter months, though not for the year as a whole ; so that after all, at least for this period of the year, it forms no exception to what is observed for the Darling Downs—that a *high admission rate* and comparatively *low mortality* are coincident with high range of temperature, low rainfall, and low amount of cloud, while a *low admission rate* and *high mortality* are associated with conditions the reverse of these.

DISEASES OF THE CIRCULATORY, NERVOUS, AND URINARY SYSTEMS.

The following notes on heart, nervous, and urinary diseases are necessarily brief, inasmuch as it is impossible to say to what extent, if at all, they are influenced by conditions of the weather; but they are of sufficient importance to call for some attention in passing. Moreover, their mean monthly curves and their rise or fall from one season to another, though not decidedly well marked, are yet sufficiently so to warrant us in assuming that these cannot possibly be a matter of mere coincidence. It will be found in the case of heart diseases (including diseases of the circulatory system generally), that the curve gradually rises from February to its maximum for the year in August, and then suddenly falls to its minimum in October; the months of *highest* death-rate being July, August, and September. The mortality curve of nervous diseases, on the other hand, shows a high maximum in February and March, falls pretty steadily to its minimum for the year in August and September, and rises again during the months of summer to its maximum the following month of February; the months of lowest death-rate being July, August, and September. It will thus be seen that these two curves are pretty well opposed to each other, and that the months of highest mortality for one are those of *lowest* death-rate for the other. The former shows a higher mortality during and immediately after the coldest months, and the latter during and immediately after the hottest months of the year. The former corresponds closely to the curve of respiratory diseases and may to some extent be due to these, the latter corresponds to the curve of gastro-intestinal diseases and may (especially in children) be intimately connected therewith.

In the case of urinary diseases the curve is somewhat irregular, but at least it will be noticed that the mortality rises in August and continues high till November.

The following tables will show the average death-rate (1887-91) from heart, nervous, and urinary diseases, for Queensland as a whole, for each month of the year according to population, and the ratio per cent. during four periods of the year :—

MORTALITY PER 100,000 POPULATION PER ANNUM.

—	Jan.	Feb.	Mar.	April.	May.	June.	July.	Aug.	Sept.	Oct.	Nov.	Dec.
Heart diseases ...	75·5	62·4	65·5	72·5	78·6	78·6	91·5	98·4	88·5	59·4	72·5	65·4
Nervous diseases	167·5	177·3	177·3	152·7	158·5	146·8	140·8	140·5	137·6	155·8	161·5	164·2
Urinary diseases	29·3	23·5	29 3	29·3	29·3	23·5	26·2	35·8	32·5	32·5	35·6	23·5

RATIO PER CENT. for FOUR PERIODS of the YEAR.

—	Jan. to Mar.	April to June.	July to Sept.	Oct. to Dec.	Total.
Heart diseases, &c. ...	22·4	25·2	30·8	21·6	100
Nervous diseases ...	27·4	25·5	21·9	25·2	100
Urinary diseases ...	23·4	23·4	27·1	26·1	100

According to these figures the difference in mortality from one month or quarter to another is not great, but it is sufficient to suggest the possibility that diseases of the circulatory, nervous, and urinary systems may at least indirectly be associated with seasonal conditions of the atmosphere.

DIARRHŒA AND DYSENTERY.

Diarrhœa and Dysentery—Queensland as a whole.—The mean monthly mortality curve for diarrhœa and dysentery for the years 1887-91 for the whole of Queensland is high in January, falls pretty regularly to its minimum for the year in August, rises somewhat from August to September, and very rapidly from September to its maximum for the year in November, and then declines a little in December. It will thus be seen that it corresponds fairly to the curves of temperature, rainfall, number of rainy days, and amount of cloud, and inversely to that of the barometer.

It must be noted, however, that the death-rate does not begin to rise till September, that it increases then much more rapidly than either the temperature, rainfall, or cloud, and attains its maximum not in the summer months of greatest heat, but in November. If the mortality from diarrhœa and dysentery be considered separately, it will be found that the curve of the former follows pretty closely that above described for diarrhœa and dysentery combined, whereas that of the latter is somewhat different, as will be seen from the following tables:—

MORTALITY PER 100,000 POPULATION PER ANNUM.

—	Jan.	Feb.	Mar.	April.	May.	June.	July.	Aug.	Sept.	Oct.	Nov.	Dec.
Diarrhœa and dysentery combined	210	172	151	157	114	88	72	57	80	184	228	183
Diarrhœa ...	126	108	98	111	78	45	48	36	58	132	168	138
Dysentery ...	84	64	53	46	36	43	24	21	22	52	60	45

RATIO PER CENT. during the FOUR QUARTERS of the YEAR.

—	Jan. to Mar.	April to June.	July to Sept.	Oct. to Dec.	Total.
Diarrhœa	29·0	20·5	12·4	38·1	100
Dysentery	36·7	22·6	12·1	28·6	100

These figures show that in the second and third quarters of the year the ratio of deaths from diarrhœa and dysentery is nearly the same, and that the former attains its maximum during the last quarter, and the latter during the first quarter of the year. This rather important difference between these two diseases may be accounted for in one or two ways. In the case of diarrhœa we find for the whole of Queensland that about 85 per cent. of the deaths occur in children under five years of age, that children are probably more susceptible than adults to the influence of a change in the weather, such as obtains in the early warm months of summer, and that it is of a more acute nature. Dysentery, on the other hand, is a disease chiefly of adult life, the percentage of deaths in children under five years of age being only 29, and it runs generally a more chronic and prolonged course.

The following will give the relative mortality of each according
to population from one year to another:—

MORTALITY PER 100,000 POPULATION for the YEARS as a Whole.[1]

—	1887.	1888.	1889.	1890.	1891.
Diarrhœa	94·0	106·7	121·5	84·3	76·4
Dysentery	48·3	50·7	57·2	31·1	39·6

It will here be observed that they rise and fall together pretty
regularly, the year of highest mortality for diarrhœa—1889—being also
the highest for dysentery, and the year of lowest mortality for one—
1891—also the lowest for the other.

The atmospheric conditions connected with the above are as
follow:—

MEAN MONTHLY METEOROLOGICAL OBSERVATIONS for the YEARS as a Whole.

—	1887.	1888.	1889.	1890.	1891.
Barometer	30·016	30·074	30·055	29·986	30·031
Temperature	70·1°	71·7°	73·5°	71·7°	70·4°
Mean range	19·4°	21·6°	21·2°	19·9°	20·7°
Absolute range	35·4°	42·7°	42·0°	38·1°	37·8°
Humidity	71%	61%	63%	67%	63%
Rainfall	3·979	2·222	2·369	4·333	3·742
Rainy days	9	5	7	9	8
Cloud	3·9	2·9	3·7	4·2	4·1
Wind	11½	12½	12	12½	14½

In comparing these meteorological readings with the mortality
statistics for the years as a whole, one thing is particularly well
marked—namely, that the year of highest death-rate has a higher mean
temperature than any other year, while those showing the lowest
death-rate have the lowest mean temperature. It would also appear
that a low mortality is associated with low barometric pressure, low
mean and absolute range of temperature, high rainfall, and high amount
of cloud, as observed in 1887, 1890, and 1891, while the reverse
of these apply to the years of highest mortality, 1888 and 1889. The
same remarks, however, that will be made in connection with typhoid
fever may also be noted here—namely, if a high barometer, high mean
and absolute range of temperature, together with low humidity and low
rainfall, contribute towards a high mortality, the rate should be higher
in 1888 than in 1889, and thus opening up the question, whether really
the death-rate in the latter year is not intensified by the conditions
that certainly exist during the year previous. Without meantime

---

[1] According to Davidson the mortality in Queensland, 1883-88, from diarrhœa,
dysentery, enteritis, and *cholera nostras* per 10,000 population is 11·43, 11·58, 3·9, and
0·93 respectively, or a total death-rate of 27·87. This is in excess of what I find, not
only for the years 1887-91, but for the longer period of twelve years, 1880-91. In the
former case the mortality from diarrhœa is 9·66, dysentery 4·54, enteritis 4·05, and
cholera nostras 0·79, or a total rate of 19·04 ; while in the latter the numbers are 10·44,
9·51, 3·29, and 0·64, or a total death-rate of 23·88 per 10,000 population per annum.
In England the rate for diarrhœa and dysentery combined (1886-90) is 6·67, or about
one-half that of Queensland, in proportion to population.

endorsing the opinion that typhoid fever and diarrhœa are caused by or associated with similar atmospheric conditions, the following table will show the ratio which they bear to themselves and towards one another, for the various years as a whole:—

RATIO PER CENT. for the YEARS 1887-91 as a Whole.

—	1887.	1888.	1889.	1890.	1891.	Total.
Typhoid fever	19·0	20·1	36·4	12·5	12·0	100
Diarrhœa	16·3	21·0	26·8	20·4	15·5	100
Dysentery	20·4	21·2	23·4	17·0	18·0	100

This will show that the mortality from typhoid fever varies much more from one year to another than either diarrhœa or dysentery, the absolute range per cent. during these five years being 24·4, 11·3, and 5·4 respectively. These figures provide fairly conclusive evidence that they are probably connected with and influenced by changes in the weather in a similar ratio, or at least that the above is the relative order of their susceptibility.

In examining the mean monthly mortality tracing for the whole period of five years, we observe a very sharp and decided rise in November, 1887, a very high and wide maximum curve embracing the months from October, 1888, to the end of May, 1889, and a comparatively low maximum curve for the spring and summer months of 1889 and 1890. In each case the death-rate begins to rise in the last quarter, and the following will show its relative magnitude from one year to another.

MEAN MONTHLY NUMBER of DEATHS—OCTOBER, NOVEMBER, and DECEMBER.

1887.	1888.	1889.	1890	1891.
75	80	53	51	49

or, according to population—

MORTALITY—OCTOBER, NOVEMBER, and DECEMBER—PER 100,000 POPULATION PER ANNUM.

1887.	1888.	1899.	1890.	1891.
266	273	174	162	148

It will here be seen that the rate during the last three years is almost equally low, and consequently their mean may be taken as representing a low mortality, and compared in one column with the other two representing both a high mortality.

MORTALITY—OCTOBER, NOVEMBER, and DECEMBER—PER 100,000 POPULATION PER ANNUM.

1887.	1888.	Mean, 1889-91.
266	273	161

MEAN MONTHLY METEOROLOGICAL OBSERVATIONS, SEPTEMBER, OCTOBER, and NOVEMBER.

—	1887.	1888.	Mean, 1889-91.
Barometer	30·031	30·099	30·019
Temperature	70·8°	74·9°	73·8°
Mean range	23·3°	25·7°	23·6°
Absolute range	42·4°	47·3°	43°
Humidity	63%	52%	56%
Rainfall	1·511	·691	1·849
Rainy days	7	3	5
Cloud...	3·5	2·5	3·2
Wind	12	12½	13½

If comparison be made between the last two columns, it is fairly evident that the high mortality of 1888 is associated with high barometric pressure, mean temperature, and mean and absolute range of temperature, together with low humidity, rainfall, number of rainy days, and low amount of cloud, while the low mortality for 1889-91 is equally associated with the reverse of these. In these respects they correspond to the atmospheric conditions observed in connection with a high and low death-rate from typhoid fever, with this exception, that the temperature is not relatively so high, in the case of a high death-rate from diarrhœa, as it is from typhoid fever. But if we compare the atmospheric conditions that obtain during the period of high mortality from diarrhœa in 1888, with those in 1887 when the death-rate is almost equally high—indeed, in one month, November, is very much higher—we shall find that they do not in any way correspond. In the one case there is high barometric pressure and high temperature, low relative humidity, and rainfall, and low amount of cloud, and in the other a specially low temperature, high humidity, high rainfall, and high amount of cloud.

Not only, therefore, is it possible to have a high mortality from diarrhœa associated with a low temperature, but we have here an instance of the rate in the month of November, 1887, rising far beyond that of any other month, while the temperature is not only low, but lower than that for the same month of any other year under review. Moreover, the temperature during the previous winter is also unusually low. It may here be noticed that the month of lowest temperature for the year is not July, as observed in the other four years, but June, the temperature for that month being 56·8, while the mean for the other years combined is 61·0; and that the temperature in July, though higher than that in June, is still comparatively low.

If, again, we examine the mortality curve for the first few months of each year, we shall find, that as a continuation of the high death-rate that existed during the last quarter of 1888 it remains high from January to the end of May, 1889, whereas in other years it is not only comparatively low, but falls either in January or February.

MEAN MONTHLY NUMBER OF DEATHS from JANUARY to MAY.

1889.		Mean for Other Years.
76	...	42

or at the rate of 252 and 139 per 100,000 population per annum respectively.

MEAN MONTHLY METEOROLOGICAL OBSERVATIONS from JANUARY to MAY.

	1889.	Mean of Other Years.
Barometer	30·031	29·971
Temperature	78·8°	75·6°
Mean range	20·8°	17·1°
Absolute range	41·4°	34·1°
Humidity	60%	69%
Rainfall	2·796	5·801
Rainy days	8	11
Cloud	3·9	4·7

The above table will show very plainly that the period of high mortality during the first few months of 1889 is connected with high barometric pressure, high mean temperature, and high mean and absolute range of temperature, together with low relative humidity, low rainfall and number of rainy days, and low amount of cloud. These observations are simply the mean of five months, but the same applies also when the comparison is made from one month to another, and it is but fair to suppose that the atmospheric conditions that favour the development of diarrhœa, as observed over an extended period of time, may also do so over a short period, and may raise or lower the mortality, if not from day to day, at least from one week to another, according to the readings of the barometer and dry and wet bulb thermometers, and to the amount of rainfall and cloud, and number of rainy days.

Diarrhœa and Dysentery combined—Moreton Division (Brisbane).—The mean monthly mortality curve for this district for the five years 1887-1891 begins in January comparatively low, and with the exception of a slight rise in April falls slowly and steadily to August, rises somewhat from August to September, and rapidly from September to October, attaining its maximum for the year in November, and falls somewhat in December—the rate for that month being a little above that of the following January. The curve showing the number of admissions to hospitals follows pretty nearly the same course, with this exception, that there is here a slight depression in February, and the maximum rate for the year is observed not in November but in October. This is, however, what might be expected, as the high admission rate for October, on account of the probable duration of the illness, naturally coincides not so much with the mortality of this month as with that of the month following. It must be remembered, also, that 84·7 per cent. of the deaths from diarrhœa in this district occur in children under five years of age, and that if these go to hospital they do so mainly as out-patients, and as they do not appear as hospital admissions, the wonder is that the mortality and admission curves so closely resemble each other.

In comparing these mean monthly curves with the various meteorological readings there does not at first sight appear to be much resemblance between the two. It will be noticed, however, that the period of low mortality for the year corresponds very fairly to the period of high relative humidity, while the two months of highest mortality— October and November—agree exactly with the two months of lowest relative humidity. Moreover, both the mean daily and absolute range of temperature have attained their maximum point in October, while the mean temperature is rapidly rising. It is true that, for two or three months previous, the mean and absolute range are both high, showing that mere range of temperature is in itself not unfavourable, as long as the mean general temperature remains comparatively low, but now that the latter rapidly rises, and that the range still continues high—the temperature being very high some days, and comparatively low other days, there being also a well-marked difference between night and day temperature—it is possible, especially after the low temperature in winter, that the effect of these sudden changes directly upon the nervous system, particularly of children, who are more susceptible than adults, is greater even than in the hot months of summer, and that in consequence the mortality rises during these months to its maximum for the year.

It is well to bear in mind also, that this is the season of the year of electric disturbances and thunderstorms, and it is pretty certain that the frequent change from positive to negative electricity and *vice versâ,* and the condition of the atmosphere generally, is more favourable to the development of certain changes in the food of children than in the winter months, thus helping indirectly to increase the mortality from diarrhœa. These changes are, however, likely, perhaps even more likely, to occur in December, January, and February, in consequence of the higher temperature of these months, and as the death-rate then declines, it would appear as if the atmosphere contributed more directly than indirectly towards this high maximum in October and November. It is also probable that the weak and ill-fed are those that succumb in late spring and early summer, and, on the principle of the survival of the fittest, those that remain are comparatively able to successfully resist the depressing influence of the excessive heat of summer, and thus reducing the mortality for the latter period below that of spring.

The following will show the mortality and admission rate to hospitals according to population, for each month of the year, based on an average of five years :—

MORTALITY and ADMISSION RATE PER 100,000 POPULATION PER ANNUM.

—	Jan.	Feb.	Mar.	April.	May.	June.	July.	Aug.	Sept.	Oct.	Nov.	Dec.
Mortality ...	204	180	156	196	134	97	84	64	134	372	444	252
Admissions ...	69	39	61	61	50	30	20	30	30	80	61	61

It will be seen from the above, that the number of admissions to hospitals must necessarily form but a trifling proportion to total number of cases that are treated either as outdoor patients or in private practice, and that therefore the value of any comparison between hospital admissions and mortality, as we shall find in the case of typhoid fever, is considerably reduced.

The relative proportion of deaths and admissions for each quarter may here be given :—

RATIO PER CENT. during the FOUR QUARTERS of the YEAR.

—	Jan. to Mar.	April to June.	July to Sept.	Oct. to Dec.	Total.
Mortality	23·3	18·4	12·2	46·1	100
Admissions	28·6	23·8	13·5	34·1	100

This will show that about one-third of the total admissions and nearly one-half of the total deaths for twelve months take place in the last quarter of the year. Again, if we take the months of lowest mortality and admission rate—June, July, and August—and compare these with those of October, November, and December, we shall find that there are 44 deaths the latter period for every 10 deaths the former, while the admission rate is 25 to 10 respectively. It would thus appear that the mortality is relatively greater than the admission rate in the last quarter, and this is what might be expected, when it is remembered that in the latter case we are dealing chiefly with adults, and in the former case with children, who, as we have seen, are highly susceptible to the influence of a rising temperature, directly and indirectly, during this period of the year.

The following tables will show the mortality and admission rate, together with the meteorological observations, from one year to another :—

MORTALITY and ADMISSION RATE PER 100,000 POPULATION for the YEARS as a Whole.

—	1887.	1888.	1889.	1890.	1891.
Mortality...	226·6	218·4	214·7	190·8	127·3
Admissions	55·4	51·2	71·6	48·9	33·5

According to these figures the general mortality is about equally high in 1887, 1888, and 1889, and lowest in 1891 ; the year of highest number of admissions being 1889, and of lowest, 1891.

MEAN MONTHLY METEOROLOGICAL OBSERVATIONS for the YEARS as a Whole.

—	1887.	1888.	1889.	1890.	1891.
Barometer	30·036	30·128	30·077	30·039	30·071
Temperature	66·7°	68·6°	69·6°	68·7°	68·4°
Mean range	16·8°	19·1°	17·4°	17·5°	17·6°
Absolute range	34·1°	36·2°	35·0°	32·5°	32·7°
Humidity	72°/₀	66°/₀	69°/₀	72°/₀	69°/₀
Rainfall	7·150	2·960	4·115	5·160	4·030
Rainy days	20	13	12	14	12
Cloud	5·0	4·5	5·5	5·7	5·1

In comparing these with each other it is quite impossible to come to any conclusion regarding the probable connection between mortality statistics and meteorological readings for the years as a whole. If we take the year showing both lowest death-rate and lowest number of

F

admissions, 1891, it cannot be said that in any respect is the cause well defined. The temperature, barometric pressure, rainfall, and humidity all occupy a medium position, and though the number of rainy days is low, it is equally low in 1889 when the mortality is higher. During the first three years under notice, the mortality, though slightly higher in 1887, is pretty nearly equal, and yet while for that year the temperature is *low*— lower than that of any other year—and the rainfall, relative humidity, and number of rainy days high, in 1889 the temperature is *high*— higher than that of any other year—the rainfall and relative humidity medium, and number of rainy days low. We have here, therefore, two years of almost equally high death-rate, the temperature being unusually high in the one case and equally low the other, showing that mere temperature in itself, whether high or low, does not, at least for *the years as a whole*, bear a constant relation to the mortality from diarrhœa.

When the monthly curve is examined from one year to another there will be observed a general rise each year in October and November, and an examination of the following figures will show that in point of mortality this is highest in 1887 and lowest in 1891.

TABLE SHOWING the DEATH-RATE for OCTOBER, NOVEMBER, and DECEMBER PER 100,000 POPULATION PER ANNUM.

—	1887.	1988.	1889.	1890.	1891.
Mortality	485	432	360	324	216
Admissions	88	72	96	48	28

MEAN MONTHLY METEOROLOGICAL OBSERVATIONS—SEPTEMBER, OCTOBER, and NOVEMBER.

—	1887.	1888.	1889.	1890.	1891.
Barometer	30·034	30·130	30·064	30·003	30·102
Temperature	67·1°	70·1°	69·4°	70·3°	68·0°
Mean range	20·0°	20·6°	18·7°	20·9°	19·6°
Absolute range	37·3°	39·8°	37·8°	41·0°	36·9°
Humidity	65%	58%	66%	59%	63%
Rainfall	3·238	2·166	5·313	2·920	3·225
Rainy days	23	9	13	8	11
Cloud	4·5	4·5	5·9	4·2	4·9
Wind	12	7	9½	8½	8

An accurate and faithful interpretation of these tables is not altogether an easy matter. If the meteorological readings associated with the low mortality in October, November, and December, 1891, be compared to those of the three previous years, when the mortality is higher, it will be observed at least that in the former year the mean temperature in the shade and absolute range of temperature are both low. If, on the other hand, we compare these with those of 1887, the year of highest death-rate for these months, we find the mean temperature lower still, while both the mean daily and absolute range of temperature are comparatively low, thus showing very conclusively that mere temperature, or its absolute range in the shade, do not bear in themselves the same relative connection to diarrhœa as is generally attributed to them. One thing may be here noted that does not appear in the above figures—namely, that the maximum radiation temperature as observed by the black bulb thermometer in the sun's rays is somewhat greater and the minimum temperature at night lower during the months September, October, and November, 1887, than any other year.

MAXIMUM TEMPERATURE in the SUN'S RAYS.

—	1887.	Mean, 1888-91.	1891.
September	143·1°	139·9°	135·1°
October	153·4°	153·2°	146·5°
November	157·0°	153·5°	151·8°

MINIMUM TEMPERATURE at NIGHT.

—	1887.	Mean, 1888-91.	1891.
September	33·4°	31·1°	36·5°
October	38·2°	39·4°	43·4°
November	48°	50°	51°

MEAN RADIATION TEMPERATURE—SEPTEMBER, OCTOBER, and NOVEMBER.

—	1887.	Mean, 1889-91.	1891.
Mean maximum	151·2°	148·9°	144·5°
Mean minimum	39·9°	40·8°	43·6°
Absolute range	111·3°	108·1°	100·9°

There can be very little doubt from these figures that, although the relative humidity is high, and, as we shall see, the number of rainy days is specially great during these months as a whole in 1887, the diathermancy of the air must on certain days have been particularly high, so as to allow of greater solar and terrestrial radiation, and it is possible that the effect of this may be all the greater during a period when the mean temperature is generally low. As regards the rainfall and number of rainy days in 1887, it will be observed that the former is medium and the latter is very high, the average number of rainy days per month in September, October, and November being 23, while the mean for the other four years is only 10. It follows, therefore, that the rain must have fallen in very gentle showers for an hour or two or three hours nearly every day, and thus afforded a not over-abundant but almost daily supply of moisture. While noting in passing that this condition may be favourable to the development of germs probably associated with these infantile disorders, it is certain at least that evaporation goes on keenly, the elastic force of vapour in the air is increased, and in consequence the barometer rises. As the mean temperature is, however, low, the rise in absolute and relative humidity soon results in further condensation of the vapour in the air, and in consequence the barometer falls. There is necessarily, therefore, in the months of September, October, and November, 1887, a greater variation in the amount of evaporation, and more constant and decided fluctuations of the barometer up and down, than observed during any other year.

During the same months, 1891, when the mortality is particularly low, there is the same total amount of rainfall, but there are on an average only 11 rainy days per month. That is to say, when it does rain, the fall must be heavier, perhaps even sufficiently so to do what the gentler showers of 1887 can probably not do—act the part of scavengers to our streets and drains, and so purify the atmosphere and our surroundings generally. Moreover, there is not the same constant variation in the amount of evaporation and fluctuation in the barometer as during a period when the rain falls for a short time nearly every day. But although a large number of rainy days is associated with a high mortality from diarrhœa in the spring months of 1887, it does not follow that the mortality will be low in proportion to the low number of rainy days, for in 1888 both the rainfall and number of wet days are unusually low, and yet the death-rate is comparatively high. If, however, we do not have in this instance to bear the effect of constant variations in evaporation and fluctuations in the barometer attending a period of many partially rainy days, we have, either as cause or effect of the opposite condition, an unusually high temperature, to contend against.

It will thus be seen that the two primary factors in connection with a high death-rate from diarrhœa are rainfall and temperature, and it seems to me that, of the two, the rainfall under certain circumstances has a closer connection than the other. A few possible reasons have already been given why the highest mortality occurs in October and November instead of in the warmer months of December and January; and moreover, if the temperature is so intimately connected with diarrhœa as is generally believed there could not possibly be a falling-off in mortality in the latter two months to the same extent as is observed. If we take the average mortality for the five years 1887-1891, we find that in October and November there are for the district of Moreton 40 deaths per month from diarrhœa, whereas in December and January there are only 22, or a little over one-half. We have seen also that the highest mortality for the last five years occurs during a season when the temperature is particularly low. I do not say that a high temperature combined with low humidity and low rainfall may not favour the development of diarrhœa—indeed it is pretty certain that under these conditions it does do so—but at the same time it is apparent that *in the presence of a medium amount of rainfall and large number of rainy days* a high summer temperature is not required to raise the mortality even to a high maximum.

One other point, regarding the force and direction of the wind and number of calms observed, may here be noticed in connection with the mortality at this season from one year to another. Bearing in mind that the death-rate is high in 1887, medium in the remaining four years as a whole, and particularly low in 1891, the following table will show its relation to north and east winds for the month of November :—

PROPORTION of NORTH and EAST WINDS for the MONTH of NOVEMBER.

—						Times Observed.	Per cent. of Whole.
N. {1887	...	...	...	...	...	14	20
{1888-91	...	...	...	...	...	8	12
{1891	...	...	...	...	...	4	7
E. {1887	...	...	...	...	...	13	18
{1888-91	...	...	...	...	...	19	31
{1891	...	...	...	...	...	19	33

PROPORTION of CALMS OBSERVED, and FORCE of WIND for the MONTH of NOVEMBER.

—						Times Observed.	Miles per Hour.
1887	...	...	...	...	...	20	14
1888-91	...	...	...	...	...	26	8
1891	...	...	...	...	...	32	$7\frac{1}{2}$

It would·appear from these figures that a period of high mortality from diarrhœa in the month of November is associated with an increased frequency of north winds, fewer number of calms, and greater velocity of wind ; and that a low mortality is connected with conditions of the wind in its direction and force pretty much the reverse of these.

REMARKS ON DIARRHŒA AND DYSENTERY.

1. That for the whole of Queensland 85 per cent. of the deaths from diarrhœa occur in children under five years of age.

2. That for the whole of Queensland 29 per cent. of the deaths from dysentery occur in children under five years of age.

3. That the highest number of *admissions to hospitals* for diarrhœa, dysentery, and enteritis takes place in the fourth quarter of the year.

4. That the highest *mortality* from diarrhœa takes place in the fourth quarter, and from dysentery in the first quarter of the year.

5. That the period of highest admission rate for dysentery and highest mortality and admission rate for diarrhœa is not coincident with the period of highest temperature, but rather with the onset of the warm season during the fourth quarter of the year.

6. That this is probably due to the sudden rise in temperature following the cold of the previous winter, as well as to its very high range at this season of the year.

7. That thunderstorms are observed at this time more frequently than during any other season of the year, and it is highly probable that the sudden change from negative to positive electricity and *vice versâ*, attending these, is associated with the high death-rate from diarrhœa.

8. That the rising and high temperature, together with the electrical condition of the atmosphere, probably act both directly upon the nervous system and indirectly through the food supply in raising the mortality and admission rate at this time of the year.

9. That the influence of temperature in its relation to diarrhœa during the *fourth* quarter of the year is not a constant quantity, and depends materially upon whether the air be humid or dry.

10. That, for the eastern portion of South and Central Queensland, a high admission rate to hospitals and high mortality from diarrhœa during the *fourth* quarter of the year are associated with a low temperature, provided it is accompanied by moderate rainfall, and either high relative humidity, or very large number of rainy days. If these latter are, however, absent, a low temperature is attended by low death-rate for that period. The higher the temperature, *if the air be dry*, the greater will be the mortality ; the lower the temperature, *if the air be humid*, the greater will be the death-rate.

11. That for the western and northern portions of the colony a high *admission rate to hospitals* during the fourth quarter of the year is associated with conditions similar to those observed for the eastern and southern portion, whereas a high *mortality* is generally connected with a high mean temperature, whether attended by high or low rainfall and relative humidity.

12. That for the western portion of the colony the type of gastro-enteric diseases attending a period of low temperature and high humidity is generally *mild ;* whereas the type of those coincident with a period of high temperature, low rainfall, and low humidity is more *malignant.*

13. That for the whole of Queensland a high mortality and admission rate during the *first* quarter of the year, are coincident with a high barometer, high temperature and high mean daily and absolute range of temperature, together with low relative humidity, rainfall, number of wet days, and low amount of cloud.

14. That the mortality and admission rate in the *first* quarter will be all the greater, the higher the barometer, mean temperature, and absolute range of temperature, and the lower the relative humidity, during the last few months of the preceding year.

15. That, generally, the higher the death-rate in the last quarter of the year (if dependent upon low temperature and high relative humidity or large number of rainy days), the *lower* it will be for the first few months of the succeeding year, and *vice versâ.*

16. That, generally, the higher the mortality in the last quarter of the year (if dependent upon high temperature and low relative humidity), the *greater* will be the mortality for the first few months of the year following.

17. That for the eastern portion of South Queensland (Brisbane) the highest death-rate for the last five years, during the fourth quarter of the year, was coincident with a greater frequency of north and fewer number of east winds than observed any other year for the same period.

18. That the influence of the velocity of the wind is doubtful, although it would appear that a high mortality is generally associated with low force of the wind.

GASTRO-INTESTINAL DISEASES (INCLUDING DIARRHŒA AND DYSENTERY).

Diarrhœa and Dysentery combined—Rockhampton Division.—The mean monthly curve of mortality for this district for the years 1887-91 begins with a high maximum in January, falls decidedly from January to February, and slowly from March to its minimum for the year in August and September, rises in October and also to its second maximum in November, and falls somewhat again in December; the months of highest death-rate being January and November, and of lowest August and September.

The following tables will give the mortality according to population for each month of the year, together with the ratio of deaths from one quarter to another :—

MORTALITY PER 100,000 POPULATION PER ANNUM (1887-91) for EACH MONTH of the YEAR.

Jan.	Feb.	March.	April.	May.	June.	July.	Aug.	Sept.	Oct.	Nov.	Dec.
444	237	264	180	108	79	108	36	43	237	357	204

RATIO PER CENT. during the FOUR QUARTERS of the YEAR.

January to March.	April to June.	July to September.	October to Dec.
41·1	16·0	8·1	34·8

This will show that for every death that takes place from July to September there are 5 from January to March, and 4 from October to December. This higher mortality during the first quarter as compared to the last quarter of the year is probably due to a larger proportion of the deaths—possibly from dysentery—occurring in adults than is observed in connection with the death-rate from diarrhœa and dysentery in the Brisbane district. We have seen that a great number of the deaths that take place in October and November are of children under five years of age; and as these form but 52 per cent. of the total deaths for the year in this district, as compared to 84·7 per cent. in Brisbane, the relative mortality for this period will in consequence be considerably reduced, while that for the months of January and February will be proportionately increased. As an instance, it may be noted here that the rate for January, 1891, is particularly high, there being recorded for that month 30 deaths out of a total of 67 for the whole year; and as only 22·5 per cent. of these were of children under five years of age, or in other words 15 in all, while 67 per cent. were of those between fifteen and twenty-five years of age, it must follow that the latter and not the former contribute chiefly towards this unusually high mortality in 1891.

The death-rate from one year to another, and the meteorological observations connected therewith, are as follow : —

MORTALITY PER 100,000 POPULATION for the YEARS as a Whole.

1887.	1888.	1889.	1890.	1891.
168	153	270	170	192

MEAN MONTHLY METEOROLOGICAL OBSERVATIONS for the YEARS as a Whole.

—	1887.	1888.	1889.	1890.	1891.
Temperature	71·2°	71·3°	72·4°	72·1°	71·6°
Mean range	13·9°	14·7°	13·9°	17·7°	18·9°
Absolute range	28·4°	29·4°	29·0°	34·4°	34·1°
Humidity	75%	74%	74%	71%	67%
Rainfall ...	3·833	3·883	2·986	6·443	4·256
Rainy days	10	6	10	13	10
Cloud	4·1	3·8	4·7	5·1	4·8
Wind	11½	13	12	12½	12

It will here be observed that the year of highest mortality, 1889, is associated with a high temperature and low rainfall, and possibly also with a low mean and absolute range of temperature. The remaining elements are more or less of a medium character.

In examining the monthly tracing from one year to another we notice that the curve rises unusually high in October and November, 1887, and January, February, and March, 1889, and specially so in January, 1891. The following tables will show the relative death-rate for October, November, and December of each year, together with the meteorological observations corresponding to the same period :—

MORTALITY PER 100,000 POPULATION PER ANNUM—OCTOBER, NOVEMBER, and DECEMBER.

1887.	1888.	1889.	1890.	1891.
511	261	180	276	204

MEAN MONTHLY METEOROLOGICAL OBSERVATIONS—SEPTEMBER, OCTOBER, and NOVEMBER.

—	1887.	1888.	1889.	1890.	1891.
Temperature	71·4°	73·8°	73·9°	74·8°	72·7°
Mean range	16·3°	16·5°	18·1°	21·8°	22·2°
Absolute range	32·0°	30·8°	33·1°	41·2°	38·3°
Humidity	70%	67%	67%	61%	58%
Rainfall ...	1·308	·585	2·794	2·703	·860
Rainy days	4	4	10	7	6
Cloud	3·6	3·6	5·1	3·9	4·9

In this instance, as was observed for the district of Brisban and for Queensland generally, the high mortality in the months o October and November, 1887, coincides with a low temperature, while at the same time the mean and absolute range of temperature are comparatively low and the humidity relatively high. Moreover, the death-rate is so particularly high, being in the proportion of fully 2 to 1 as compared to the mean of other years, and the temperature so comparatively low, being 71·8 deg. against 73·8 deg. for other years, that when combined either with high relative humidity or (as observed for Brisbane) a large number of rainy days with moderate rainfall, a low temperature undoubtedly seems to be intimately associated with a high mortality during the last quarter of the year.

As regards the comparatively high death-rate in January, February, and March, 1889, and in January, 1891, it will be observed that in both cases, on the other hand, the temperature and absolute range of temperature are unusually high, not only at the time, but for a month or two previous, while the humidity is also in both instances relatively low. It may be specially noted that for several months in the latter part of 1890, and immediately prior to the exceedingly high mortality in January, 1891, the absolute range of temperature is exceptionally high and the relative humidity equally low. In this case at least there can be no doubt that the great bulk of the deaths occur in adults, for, as we have seen, only 15 of those for the whole year are of children under five years of age, while in this month alone, 30 deaths are recorded of all ages.

It would seem from these observations that for this district a high mortality during the *last* quarter of the year coincides with a low temperature and low absolute range of temperature, combined with high relative humidity; while a high death-rate during the *first* quarter is associated with a high temperature, high absolute range of temperature, and low relative humidity, and in the latter case is all the higher if these conditions obtain also for the last two or three months of the previous year. If that be so, it follows that if the atmospheric conditions in the last quarter of the year are favourable to an exceptionally high mortality at that time, as observed in 1887 (low temperature and low range and high humidity), the mortality in the first quarter immediately following will be all the less, as the conditions that predispose to a high death-rate at that time (high temperature and low humidity) are not present also in the latter months of the preceding year. The higher the death-rate, therefore, during the last quarter the less it will be the first quarter, and *vice versâ.* A low death-rate during the last quarter may, however, not be followed by a high mortality the first quarter if the atmospheric conditions are equally favourable to a low rate, We have seen that the highest percentage of deaths from diarrhœa takes place the last quarter, and from dysentery the first quarter, of the year. We have also seen that the majority of cases of diarrhœa occur in children under five years of age, while dysentery, on the other hand, is chiefly prevalent amongst adults. It is highly probable that the gastro-enteric diseases of adult life run a more chronic course than those of childhood, and that, although generally showing a higher mortality the first quarter, they may have been contracted the last quarter of the preceding year. In this way we have an explanation of the observation made above, that

while a high death-rate the first few months of the year is associated
with certain atmospheric conditions at the time, this rate is all the
higher if those same conditions are present during the last few months
of the previous year. It may therefore fairly be concluded that an
exceptionally low temperature, low range of temperature, and high
relative humidity (or large number of rainy days), in the months of
October, November, and December, are coincident with a large increase
in the number of cases of diarrhœa in children, which run an acute
course, and die during the last quarter of the year; and that the
opposite conditions (high temperature, and high range of temperature,
and low relative humidity) during these same months are associated
with a great increase in the number of enteric diseases in both children
and adults, which run a more or less sub-acute course and die during
the first quarter of the year following.

Gastro-Intestinal Diseases—Cooktown Division.—The mean
monthly mortality and admission curves for this district for the years
1887-91 show a high rate the first few months of the year, a decline
during the months of winter, and a slight tendency to rise the last
quarter of the year.

RATIO PER CENT. during the FOUR PERIODS of the YEAR.

—	Jan. to Mar.	April to June.	July to Sept.	Oct. to Dec.
Admissions	38·4	22·3	15·6	23·7
Mortality	30·5	24·0	20·3	25·2

According to these figures there is a slightly greater difference
between the summer and winter months in the number of admissions
than there is in the death-rate, and it follows if the same proportion
of cases are regularly sent to hospital either that the type of the disease
is more severe in the cold months, and relatively to the number of
admissions the mortality is high, or that those illnesses contracted
during the first quarter are sub-acute in their character, and so contribute to the death-rate of July, August, and September more than
to that of the early months of the year.

The following will show the mortality and admission rate according
to population, together with the meteorological observations for the
years as a whole:—

MORTALITY and ADMISSION RATE PER 100,000 POPULATION for the YEARS as
a Whole.

—	1887.	1888.	1889.	1890.	1891.
Admissions	348	439	521	205	109
Mortality	552	320	680	379	250

MEAN MONTHLY METEOROLOGICAL OBSERVATIONS for the YEARS as a Whole.

—	1887.	1888.	1889.	1890.	1891.
Temperature	77·9°	77·9°	79·6°	79·0°	77·1°
Mean range :..	13·3°	11·9°	11·2°	10·5°	9·7°
Absolute range	23·2ʳ	24·5°	22·0°	22·3°	21·5°
Humidity	73°/₀	72°/₀	73%	68°/₀	68%
Rainfall	5·526	4·981	4·366	6·167	5·758
Rainy days	11	10	9	10	11
Cloud	4·2	3·4	3·8	4·7	5·1
Wind	11½	11½	10½	11	14½

It will be observed that the year of highest mortality and admission rate (1889) corresponds to the year of highest temperature, comparatively low rainfall and number of rainy days, low amount of cloud and force of wind, while the year of lowest death-rate and lowest number of admissions (1891) coincides with atmospheric conditions pretty well the opposite of these. It may also be noticed that in the former case the relative humidity is high, and in the latter case low, and that this is the reverse of what is usually noted in connection with a high or low mortality respectively.

The above tables will also show, that while the admission rate is comparatively high the mortality is relatively low in 1888, and if there be any constant connection between these it follows that though the number of illnesses be high they run a mild course, and a large proportion of the patients recover. There does not, however, appear to be any peculiarity in the meteorological observations to warrant us in coming to any conclusion regarding the matter.

As regards the high mortality and admission rate for 1889, it will be found on examining the monthly tracing that they are pretty well distributed over the year, although both are specially high the first three months, as will be seen from the following :—

MORTALITY and ADMISSION RATE PER 100,000 POPULATION PER ANNUM—JANUARY, FEBRUARY, and MARCH.

	1889.		Mean of Other Years.
Admissions	... 709	...	434
Mortality ...	... 921	...	416

We here note that the ratio of admissions is almost, and of death-rate fully, 2 to 1 that of other years. The following are the meteorological readings coincident with the above :—

MEAN MONTHLY METEOROLOGICAL OBSERVATIONS—JANUARY, FEBRUARY, and MARCH.

				1889.		Mean of Other Years.
Temperature	...	...	...	83·6°	...	82·0°
Mean range	...	...	...	12·7°	...	12·8°
Absolute range	...	...	...	25·0°	...	23·2°
Humidity ...	...	...	...	73%	...	75%
Rainfall	...	...	...	6·030	...	13·284
Rainy days ...	...	...	...	12	...	18
Cloud	...	...	...	3·8	...	5·2

This fairly shows that the period of high mortality and admission rate in 1889 is associated with a high temperature, low rainfall, and number of rainy days, and low amount of cloud. It may be mentioned that neither here, nor when comparing one year with another as a whole, are the mean and absolute range of temperature sufficiently decided to connect them with either a high or low mortality. This is hardly to be wondered at when it is remembered that the meteorological observations for Cooktown are, with the exception of rainfall and force of wind, more uniform, not only from year to year, but from one month or season to another, than for any other station under consideration. This, of course, is due to its geographical position in the tropics.

Gastro-Intestinal Diseases—Normanton Division.—The mean monthly curves for gastro-enteric diseases for this division for the years 1888-91 show that the highest number of admissions takes place the last quarter, and the highest mortality the first quarter of the year. This is well seen in the following table :—

RATIO PER CENT. during the FOUR QUARTERS of the YEAR.

—	Jan. to Mar.	April to June.	July to Sept.	Oct. to Dec.
Admissions	24·2	27·4	13·0	35·4
Mortality	37·3	19·2	18·1	25·4

showing that the illnesses of the last quarter are probably of a subacute character, and tend to increase the death-rate, not at the time, but during the early months of the following year.

MORTALITY and ADMISSION RATE PER 100,000 POPULATION during the YEARS as a Whole.

—	1888.	1889.	1890.	1891.
Admissions	912	1,212	540	264
Mortality	1,248	1,128	780	480

It will here be observed that the mortality is high for the first two years, more especially in 1888; that the number of admissions is greatest in 1889; and that in both cases the rate is much reduced in 1891.

MEAN MONTHLY METEOROLOGICAL OBSERVATIONS for the YEARS as a Whole.

—	1888.	1889.	1890.	1891.
Barometer	30·025	29·993	29·935	29·989
Temperature	81·0°	83·1°	79·6°	77·9°
Humidity	65°/₀	69°/₀	58°/₀	58°/₀
Rainfall	3·106	1·311	5·795	4·581
Rainy days	5	4	7	5
Cloud	2	2·7	3·5	3·9

This will show that the two years of highest mortality and number of admissions correspond to years of high barometric pressure and temperature, low rainfall, number of rainy days, and low amount of cloud. Curiously enough the relative humidity is high also, this being one of the instances in Queensland where a high relative humidity and high temperature are together associated with an increased death and admission rate. As a rule these two atmospheric conditions, in their relation to gastro-intestinal diseases, are opposed to each other; that is to say, a high temperature is usually connected with a relatively low humidity, and *vice versâ*.

In examining the monthly tracing from one year to another, it will be observed that the death-rate rises very high in March, 1889.

MORTALITY and ADMISSION RATE—JANUARY, FEBRUARY, and MARCH—PER 100,000 POPULATION PER ANNUM.

—	1889.	Mean of Other Years.
Admissions	1,078	561
Mortality	2,536	1,159

It will here be seen that the number of deaths and admissions in the first quarter of 1889 is about double the average for other years.

MEAN MONTHLY METEOROLOGICAL OBSERVATIONS, DECEMBER, JANUARY, and FEBRUARY.

—	1888-89.	Mean of Other Years.
Barometer	29·939	29·805
Temperature	86·8°	83·7°
Humidity	76%	75%
Rainfall	2·813	14·639
Rainy days	8	16
Cloud	5·3	6·4

It will be noted that the increase in number of deaths and admissions for this period is connected with atmospheric conditions similar to those observed when comparing the death-rate from one year to another as a whole.

Gastro-Intestinal Diseases—Cloncurry and Hughenden Division.—The mean monthly curve of mortality for this division for the years 1888-91 shows a pretty uniform rate for the first, second, and fourth quarters of the year, and a low rate for the third, while in number of admissions the first is very high, and the last unusually low.

The following figures will show their ratio per cent. for four periods of the year:—

RATIO PER CENT. during the FOUR QUARTERS of the YEAR.

—	Jan. to March.	April to June.	July to Sept.	Oct. to Dec.
Admissions	36·2	23·9	23·0	16·9
Mortality...	29·1	27·9	14·0	29·0

The total number of deaths and admissions is, however, so very small for this large but thinly populated district, that these can at best give but an approximate idea of their relationship to one another.

MORTALITY and ADMISSION RATE PER 100,000 POPULATION for the YEARS as a Whole.

—	1888.	1889.	1890.	1891.
Admissions	216	432	48	228
Mortality	459	552	252	228

This will show that both the mortality and admission rate rise to their maximum in 1889, that in 1890 the number of admissions is very low relatively to that of other years, and to the number of deaths for that year, and that in 1891, while the admission rate is comparatively high, the death-rate is lower than that of any other year.

MEAN MONTHLY METEOROLOGICAL OBSERVATIONS for the YEARS as a Whole.

—	1888.	1889.	1890.	1891.
Temperature	76·5°	78·7°	75·5°	72·8°
Mean range	27·5°	26·2°	24·9°	24·9°
Absolute range	44·9°	48·6°	46·0°	42·7°
Humidity	49°/₀	53°/₀	59°/₀	54°/₀
Rainfall	1·861	1·336	2·168	3·474
Rainy days	3	4	6	5
Cloud	1·5	2·2	2·9	2·8
Wind	13	12	12½	12½

It will here be observed that the year of high mortality and admission rate corresponds to the year of highest temperature, high mean daily and absolute range of temperature, low rainfall, and number of rainy days, and low amount of cloud. With the exception of temperature and its absolute range and rainfall, these are, however, more extreme the previous year, and although the rate of admissions especially, is not so high during that year, it is probable that the atmospheric conditions of 1888 contribute somewhat to the high death and admission rate in 1889. It may also be noted that in 1891 there is a low temperature, high rainfall, and high amount of cloud, and that, although this is the year of lowest mortality, the admission rate is comparatively high. It would appear that while a large number of admissions may be coincident with a low temperature, the cases run a mild course and generally recover.

Gastro-Intestinal Diseases—Boulia and Blackall Division.—The mean monthly mortality and admission curves for the years 1888-91 show a high rate in the former case, for the first quarter, and in the latter case for the last quarter of the year.

RATIO PER CENT. for the FOUR QUARTERS of the YEAR.

—	Jan. to March.	April to June.	July to Sept.	Oct. to Dec.	Total.
Admissions ...	29·4	23·8	13·8	33·0	100
Mortality ...	39·3	19·4	21·6	19·7	100

It would appear from these figures either that the proportion of recoveries to number of sicknesses in the last quarter is comparatively high, thus reducing the death-rate for this period, or, what is more likely, that the diseases admitted in November and December are chiefly of a sub-acute character, and of two or three months' duration, and so tend to increase the death-rate, not at the time, but during the first few months of the year following.

MORTALITY and ADMISSION RATE PER 100,000 POPULATION for the YEARS as a Whole.

—	1838.	1889.	1890.	1891.
Admissions	348	564	708	612
Mortality	300	552	300	228

This shows that the highest mortality is recorded in 1889 and the highest number of admissions in 1890 and 1891, and if the same proportion of cases that occur in general practice are sent to hospital from one year to another it follows that the type of the disease is upon the whole mild, and that the percentage of recoveries is very much greater the last two years than during the other two. The meteorological observations are as follow :—

MEAN MONTHLY METEOROLOGICAL OBSERVATIONS for the YEARS as a Whole.

—	1888.	1889.	1890.	1891.
Temperature	74·5°	75·6°	70·8°	70·3°
Mean range	31·2°	29·5°	23·4°	25·1°
Absolute range	52·5°	56·2°	43·7°	43·9°
Humidity	47°/₀	52°/₀	57°/₀	60°/₀
Rainfall	1·311	1·151	2·599	2·503
Rainy days	3	4	5	5
Cloud	1·8	2·5	3·3	3·2
Wind	18	13½	13½	13½

It is particularly evident here that the high death-rate in 1889 is associated with a high temperature and high mean daily and absolute range of temperature, together with low relative humidity, rainfall, number of rainy days, and low amount of cloud; the force of the wind being somewhat doubtful. It will also be noticed that these conditions are present the previous year and it is probable that these helped to increase the mortality in 1889. Moreover, it should be observed that in both 1890 and 1891, when the death-rate is low, the meteorological readings are of quite an opposite character—namely, low temperature and low range of temperature, high humidity, rainfall, number of rainy days, and cloud. It must be remembered, however, that the admission rate is particularly high both of these years, and if there is any constant relation between the mortality and number of hospital admissions, it must follow that, while a low temperature and the other atmospheric conditions usually accompanying a low temperature, as observed above, contribute towards a high admission rate, they produce a mild form of the disease, and so do not bring about a corresponding increase in the mortality. Conversely, a high temperature together with its attendant atmospheric conditions is connected with a more malignant form of gastro-enteric diseases, and increases the death-rate to an extent beyond its usual ratio to the admission rate.

As regards this high mortality in 1889, it will be found on examining the monthly tracing to occur chiefly in February and March, the number of deaths for these two months being on an average 12 per month, while the mean number for other years is 3·5 or at the rate of 1,357 and 383 respectively per 100,000 population per annum. In other words, the death-rate from gastro-enteric diseases in February and March, 1889, is greater than that of the other three years combined. The following meteorological observations are coincident therewith :—

MEAN MONTHLY METEOROLOGICAL OBSERVATIONS—DECEMBER, JANUARY, FEBRUARY.

	1888-89.	Mean of Other Years.
Barometer	29·860	29·774
Temperature	89·8°	83·1°
Mean range	32·9°	24·4°
Absolute range	55·5°	41·7°
Humidity	38%	53%
Rainfall	·820	4·333
Rainy days	4	8
Cloud	2·4	4·1
Wind	12	15

It will be observed that the period of high mortality is connected with a comparatively high barometer, high temperature and mean daily and absolute range of temperature, low relative humidity, rainfall, number of rainy days, amount of cloud, and low force of wind. At the same time, there is no corresponding increase in the admission rate either at the time or during the last months of the previous year, and it would appear as if the nature of the disease is very malignant, and that a large percentage of the patients die. In the month of February, 1890, on the other hand, the admission rate is unusually high, but there is no corresponding increase in the mortality, showing that the diseases, though prevalent, are mild in their type, and that the great proportion of the patients recover. In this latter case the temperature, mean and absolute range of temperature, and barometer are low, while the humidity, rainfall, number of wet days, and amount of cloud are all high; or in other words, exactly the reverse of what is observed in connection with a period of high mortality and low admission rate. The type of the disease is in the one case mild and in the other malignant.

Gastro-Intestinal Diseases—Roma and Thargomindah Division.— The mean monthly mortality and admission curves for gastro-enteric diseases for this division for the five years 1887-1891 show a comparatively high rate during the first quarter and a pretty uniformly low rate for the remainder of the year.

RATIO PER CENT. during FOUR PERIODS of the YEAR.

—	Jan. to Mar.	April to June.	July to Sept.	Oct. to Dec.	Total.
Admissions	40·2	25·1	14·1	20·6	100
Mortality	35·9	22·3	22·9	18·9	100

The following will show the relative proportion of admissions and mortality from one year to another as a whole :—

MEAN ADMISSION RATE and MORTALITY PER 100,000 POPULATION for the YEARS as a Whole.

—	1887.	1888.	1889.	1890.	1891.
Admissions	132	156	183	152	100
Mortality	127	192	400	216	213

In both admission and death rate the year 1889 stands highest. This is more apparent in the latter case, the number of deaths being about double the average observed in other years. The following table will give for comparison the mean monthly meteorological observations for the years as a whole :—

MEAN MONTHLY METEOROLOGICAL OBSERVATIONS for the YEARS as a Whole.

—	1887.	1888.	1889.	1890.	1891.
Temperature	66·7°	67·9°	70·5°	68·1°	67·7°
Mean range	24·8°	28·2°	24·1°	22·4°	24·8°
Absolute range	46·9°	51·4°	49·4°	44·2°	44·6°
Humidity	67%	50%	61%	66%	63%
Rainfall	2·244	1·031	1·814	3·503	2·717
Rainy days	5	3	5	6	5
Cloud	2·9	2·4	4·0	3·9	3·8

It will be observed that the year of highest mortality coincides with a year of high temperature and high absolute range of temperature, low humidity, and low rainfall. It is true that with the exception of temperature these are more pronounced the preceding year of 1888, and here again giving further evidence in support of the statement made elsewhere, that diarrhœa, like typhoid fever, may be influenced by the condition of the weather that obtains for several months previous to and during the period of high mortality. This

G

period of high death and admission rate begins in December, 1888,
rises to its maximum in February, and continues more or less high
during the greater part of 1889.

The relative magnitude of the mortality for this period as compared to that of other years may be seen from the following table :—

MORTALITY and ADMISSION RATE PER 100,000 POPULATION PER ANNUM during
the Months of DECEMBER, JANUARY, and FEBRUARY.

—	1888-89.	Mean of Other Years.
Admissions	285	157
Death rate	485	257

showing that in both admission and death rate the proportion for the
summer months of 1889 as compared to that of other years is nearly
2 to 1.

MEAN MONTHLY METEOROLOGICAL OBSERVATIONS—DECEMBER, JANUARY, and
FEBRUARY.

	1888-89.	Mean of Other Years.
Temperature	86·3°	79·3°
Mean range	28·6°	22·1°
Absolute range	56·3°	45·4°
Humidity...	47°/₀	53°/₀
Rainfall	1·755	4·421
Rainy days	3	7
Cloud	2·8	4·1
Wind	12	15

According to these figures, an unusually high death and admission
rate is coincident with a high temperature and high mean daily and
absolute range of temperature, together with low relative humidity,
rainfall, number of rainy days, and cloud, and low force of wind. As
regards the force of the wind, it should be noted that it is lower in the
summer months of 1888, the mortality being at the same time also
lower, so that it is doubtful to what extent the mere velocity of the
wind is associated with mortality. It may be observed, however, that
from the beginning of November, 1888, to the end of May, 1889, the
prevailing wind is constantly from the east, whereas in other years it
is either north or north-east. If there is, therefore, any relation
between the two, it would appear as if for this large western district of
South Queensland the influence of the winds in their direction is
exactly opposed to what is observed for the eastern portion of the
colony, where east winds are favourable and north winds unfavourable
towards a low mortality from diarrhœa.

Gastro-Intestinal Diseases—Darling Downs.—The mean monthly
curve of admission rate to hospitals for this district for the years
1887-91 for all acute and sub-acute gastro-intestinal diseases shows
a high maximum in November, a somewhat lower rate for the first
quarter, and a low rate during the remaining months of the year. It
may be noted, however, that the mortality is low in November and

high in December, and also comparatively high during the first few months of the year. The following will give their ratio per cent. for four periods of the year :—

RATIO PER CENT. during the FOUR QUARTERS of the YEAR.

—	Jan. to Mar.	April to June.	July to Sept.	Oct. to Dec.	Total.
Admissions	31·7	18·4	18·4	31·5	100
Deaths	35·7	20·5	17·8	26·0	100

This will show that the number of deaths is relatively much greater in the first quarter than in the last quarter of the year, while the admission rate is about equal in both cases. It is probable that this difference is due either to a large proportion of the cases that are admitted during the latter part of the year being of a sub-acute nature, and thus adding to the mortality in the beginning of the year following ; or that the type of the disease is more virulent the latter period, and a greater proportion of them die than in the months of November and December.

The following table will show the relative admission and death rate from year to year as a whole :—

MEAN MORTALITY and ADMISSION RATE PER 100,000 POPULATION for the YEARS as a Whole.

—	1887.	1888.	1889.	1890.	1891.
Admission rate	228	252	294	194	166
Mortality	80	109	169	79	105

and for comparison the mean monthly meteorological readings are here added :—

MEAN MONTHLY METEOROLOGICAL OBSERVATIONS for the YEARS as a Whole.

—	1887.	1888.	1889.	1890.	1891.
Temperature	61·5°	62·8°	63·4°	62·8°	62·0°
Mean range	19·3°	24·8°	21·5°	21·7°	21·7°
Absolute range	36·5°	46·7°	44·9°	42·3°	41·5°
Humidity	79%	70%	73%	74%	74%
Rainfall	4·005	1·723	3·052	3·743	3·329
Rainy days	10	6	10	12	10
Cloud	3·2	3·5	4·5	5·1	4·5
Wind	8	10½	12	12	13½

It will be observed that the year of highest mortality and highest admission rate—1889—corresponds to a year of high temperature, high mean and absolute range of temperature, low relative humidity, and low rainfall. With the exception of temperature, these, however, are more decided the year previous, showing possibly, as has already been observed in connection with diarrhœa, and as will be noted more especially for typhoid fever, that the meteorological conditions prevalent in 1888 may have for many months prior to the period of high mortality in 1889 contributed towards the specially high rate of the latter year as a whole.

It may also be noted here that the year of lowest mortality, 1887, is associated with a low temperature and low mean and absolute range of temperature, high relative humidity, and high rainfall. At the same time the number of admissions for that year is comparatively high, and if any reliance can be placed on the relation between the hospital admission rate and the total mortality (including, of course, the deaths that take place outside the hospital), it must follow that, though prevalent in 1887, the type of the disease is mild, and a large percentage of the cases recover, and in consequence the mortality for that year is low. It would therefore seem that a high temperature and its usual atmospheric attendants not only cause a large increase in the number of cases of diarrhœa, but the nature of the disease is very severe and the death-rate is unusually high, while a low temperature and high rainfall may cause an increase in the admission rate, but the disease is of a mild type, and the cases usually recover.

In examining the monthly curve it will be found that for the period above noted the high admission rate begins in November, and the high death-rate in December, 1888, and that both continue unusually high during the greater part of 1889. The following table will give the relative magnitude of the admission and death rate for the months of December, January, and February, as compared to the mean of other years :—

MEAN MORTALITY and ADMISSION RATE PER 100,000 POPULATION PER ANNUM during the MONTHS of DECEMBER, JANUARY, and FEBRUARY.

	1888-89.	Mean of Other Years.
Admissions	584	209
Mortality	209	109

These figures show that in point of mortality it is almost double, and in admission rate almost three times that of the mean of other years. The meteorological observations associated therewith may here be given for comparison.

MEAN MONTHLY METEOROLOGICAL OBSERVATIONS—DECEMBER, JANUARY, and FEBRUARY.

	1888-89.	Mean of Other Years.
Temperature	74·7°	72·1°
Mean range	26·3°	18·2°
Absolute range	49·5°	37·1°
Humidity	67%	73%
Rainfall	2·736	6·127
Rainy days	8	14
Cloud	3·5	6·3
Wind	11	13

There is here a very distinct connection between a high mortality during the summer months, and high temperature, high mean and absolute range of temperature, low humidity, rainfall, and number of rainy days, and low amount of cloud.

TYPHOID FEVER.

Typhoid Fever—Queensland as a Whole.—The mean monthly mortality curve of typhoid fever, for the quinquennial period 1887-91, for the whole of Queensland shows a high maximum in January, February, and March, a rapid fall from March to April, a gradual decline from April to its minimum for the year in August, a low rate for September and October, followed by a sharp rise to December. This corresponds very closely to the curve of mean temperature, rainfall, number of wet days, and amount of cloud, especially the three last named.

The following will show the death-rate according to population for each month, and the ratio per cent. for four periods of the year :—

MEAN MORTALITY PER 100,000 POPULATION PER ANNUM for EACH MONTH
of the YEAR.

Jan.	Feb.	March.	April.	May.	June.	July.	Aug.	Sept.	Oct.	Nov.	Dec.
72·0	73·2	69·6	42·0	44·0	38·0	30·0	15·6	21·6	21·6	38·4	54·0

RATIO PER CENT. for the FOUR QUARTERS of the YEAR.

January to March.	April to June.	July to September.	October to December.	Total.
41·3	23·9	12·9	21·9	100

It will be observed that the months of highest mortality are January, February, and March, and of lowest death-rate August, September, and October, and that for every 10 deaths that take place the latter period there are 44 the former. If, however, the year be divided into the ordinary quarters as above tabulated, it will be found that there are 32 deaths recorded the first quarter for every 10 the third quarter of the year.

The death-rate for the years as a whole according to population is as follows :—

MORTALITY per 100,000 POPULATION for the YEARS as a Whole.[1]

1887.	1888.	1889.	1890.	1891.
40·6	46·4	77·2	27·1	26·9

[1] According to these figures the average annual death-rate from typhoid fever for the last five years per 100,000 population is 43·6. For the twelve years 1880-91 it is 64·2, the great increase here noted being due chiefly to an exceedingly high rate for the year 1884. In England the death-rate (1886-90) is 17·9, showing that for every 10 deaths in England there are 24 in Queensland in proportion to the population.

These figures show very plainly that the year of highest mortality is 1889, and that for the last two years the death-rate is particularly low. If the total number of deaths from typhoid fever in Queensland for the last five years be taken into consideration, it will be found that 35·4 per cent. of these occur in 1889, only 12·5 per cent. in 1890, and 12 per cent. in 1891, or in the proportion of about 3 in 1889 to 1 during the last two years.

For convenience the meteorological readings for the years as a whole, in connection with the above, may here be reinserted:—

MEAN MONTHLY METEOROLOGICAL OBSERVATIONS for the YEARS as a Whole.

—	1887.	1888.	1889.	1890.	1891.
Barometer	30·016	30·074	30·055	29·986	30·031
Temperature	70·1°	71·7°	73·5°	71·7°	70·4°
Mean range	19·4°	21·6°	21·2°	19·9°	20·7°
Absolute range	35·4°	42·7°	42·0°	38·1°	37·8°
Humidity	71%	61%	63%	67%	63%
Rainfall	3·979	2·222	2·369	4·333	3·742
Rainy days	9	5	7	9	8
Cloud	3·9	2·9	3·7	4·2	4·1
Wind	11½	12½	12	12½	14½

In comparing the years of high or low mortality with the corresponding meteorological observations as represented in this table, there does not appear to be much difference in these from one year to another, certainly not sufficient to account for the great difference in death-rate above noted during the last three years. It is true that if we compare the atmospheric conditions of 1889 with those of 1887, 1890, and 1891, we shall find in the former case that the barometric pressure, temperature, and absolute range of temperature are high, while the rainfall and amount of cloud are equally low ; but if on the other hand the observations of 1889 be compared to those of 1888, it will be found that with the exception of mean temperature, they are even more pronounced in favour of a higher mortality the latter year. We shall see, however, later on that this offers no exception to but rather bears testimony in support of the ordinary rule that typhoid fever is intimately associated with high temperature, low humidity, and low rainfall, and that if these conditions in 1888 do not increase the death-rate for the time being it would appear as if they very materially contribute towards the exceptionally high mortality of the succeeding summer.

In examining the monthly tracings we find that this high mortality curve begins in November, 1888, and continues to rise steadily until March, 1889, when it attains its maximum, then suddenly falls from March to April, but still continues comparatively high till the end of July, the death-rate for that month being greater—with the exception of January and February, 1887—than the maximum of any other year even during the summer months. There are thus nine months of a continuously high mortality ; a fair idea of which will be formed when compared with that of other years for the same period.

MORTALITY PER 100,000 POPULATION PER ANNUM.

—	March.	Jan., Feb., and March.	Nov. to July.
1889	168·0	132·5	98·0
Mean of other years	45·6	56·4	39·0

It will here be observed that for the month of March, 1889, the mortality is nearly four times the average of other years, or in other words almost equal to the total number of deaths that occur in the month of March for the other four years combined. If the mean mortality be taken for January, February, and March, or for the nine months from November, 1888, to July, 1889, the death-rate is nearly three, times the average of other years during the same periods.

We have already seen what the atmospheric conditions for the year 1888 as a whole are, but in order to follow closely this long period of high death-rate it may be well to give the meteorological observations for a month or two previous to its commencement, and compare these with the mean of other years.

MEAN MONTHLY METEOROLOGICAL OBSERVATIONS—SEPTEMBER, OCTOBER, and NOVEMBER.

	1888.		Mean of Other Years.
Barometer	30·099	...	30·030
Temperature	74·9°	...	73·1°
Mean range	25·7°	...	23·5°
Absolute range	47·3°	...	42·9°
Humidity	52%	...	58%
Rainfall	·691	...	1·764
Rainy days	3	...	6
Cloud	2·5	...	3·3
Wind	12½	...	13

This shows, more distinctly even than the observations for 1888 as a whole, that preparatory to the rise in mortality, the barometric pressure, mean temperature, and its mean and absolute range are high, while the relative humidity, rainfall, number of wet days, and amount of cloud are particularly and equally low.

The following are the meteorological observations for the months of December, 1888, and January and February, 1889, and which correspond to the maximum death-rate during the first three months of the latter year, and for comparison those showing the mean of other years.

MEAN MONTHLY METEOROLOGICAL OBSERVATIONS—DECEMBER, JANUARY, and FEBRUARY.

	1888-89.		Mean of Other Years.
Barometer	29·965	...	29·851
Temperature	83·3°	...	79·4°
Mean range	23·7°	...	17·4°
Absolute range ...	44·2°	...	33·8°
Humidity	50%	...	67%
Rainfall	2·366	...	7·800
Rainy days	7	...	13
Cloud	3·7	...	5·1
Wind	11	...	12½

The difference in meteorological readings is here particularly well marked, the barometer, temperature, and its mean and absolute range being on the one hand very high, and on the other comparatively low, while the relative humidity, rainfall, number of rainy days, and amount of cloud are in the one case especially low, and in the other equally high. We shall also find, if a similar comparison be made for the next three months, that the continued high mortality in April, May, and June is associated with atmospheric conditions, if not as decided as, at least pretty similar to these.

MEAN MONTHLY METEOROLOGICAL OBSERVATIONS—MARCH, APRIL, and MAY.

		1889.		Mean of Other Years.
Barometer	...	30·097	...	30·040
Temperature	...	74·4°	...	71·6°
Mean range ...	...	18·0°	...	17·2°
Absolute range	...	38·6°	...	32·6°
Humidity	...	70%	...	74%
Rainfall ...	...	3·227	...	4·729
Rainy days	...	9	...	11
Cloud ...	...	4·2	...	4·5

In the above table the year 1888 is left out in forming a mean for other years, as it has been already suggested that it either directly or indirectly tends to increase the death-rate for 1889, and probably for this reason might rather have been included with that year to form a mean for the two.

That this suggestion is absolutely correct I think there can be little doubt. We have seen that the atmospheric conditions for the latter months of 1888 are intimately connected with the high mortality of the summer following ; and as these conditions obtain for some time prior even to this, it seems highly probable that during the greater portion of that year they are equally associated with the death-rate of the following year. If the various meteorological curves for the two years previous to the summer of 1889 be carefully examined, it will be found that the barometer, temperature, and mean and absolute range of temperature have a gradual tendency upwards—that is to say, these are higher in the winter months of 1888 than 1887, and in the summer months of 1889 than 1888—while the relative humidity, rainfall, number of wet days, and amount of cloud have an equal and decided tendency downwards. One cannot, indeed, help coming to the conclusion that for a period of many months beforehand the weather is preparing the way, the atmospheric conditions becoming more and more favourable for the development of those germs that are known to be the cause of typhoid fever, and that when these conditions culminate in a very high temperature, in the absence of the normal fall of rain during the summer months of 1889, the mortality rises to an extent beyond that observed for any other period under review.

CHART (FOR QUEENSLAND AS A WHOLE),

showing that for a period of two years prior to the epidemic of typhoid fever in the summer months of 1888-89 the barometer, mean temperature, and mean and absolute range of temperature have a gradual tendency upwards, while the relative humidity has an equal and decided tendency downwards.

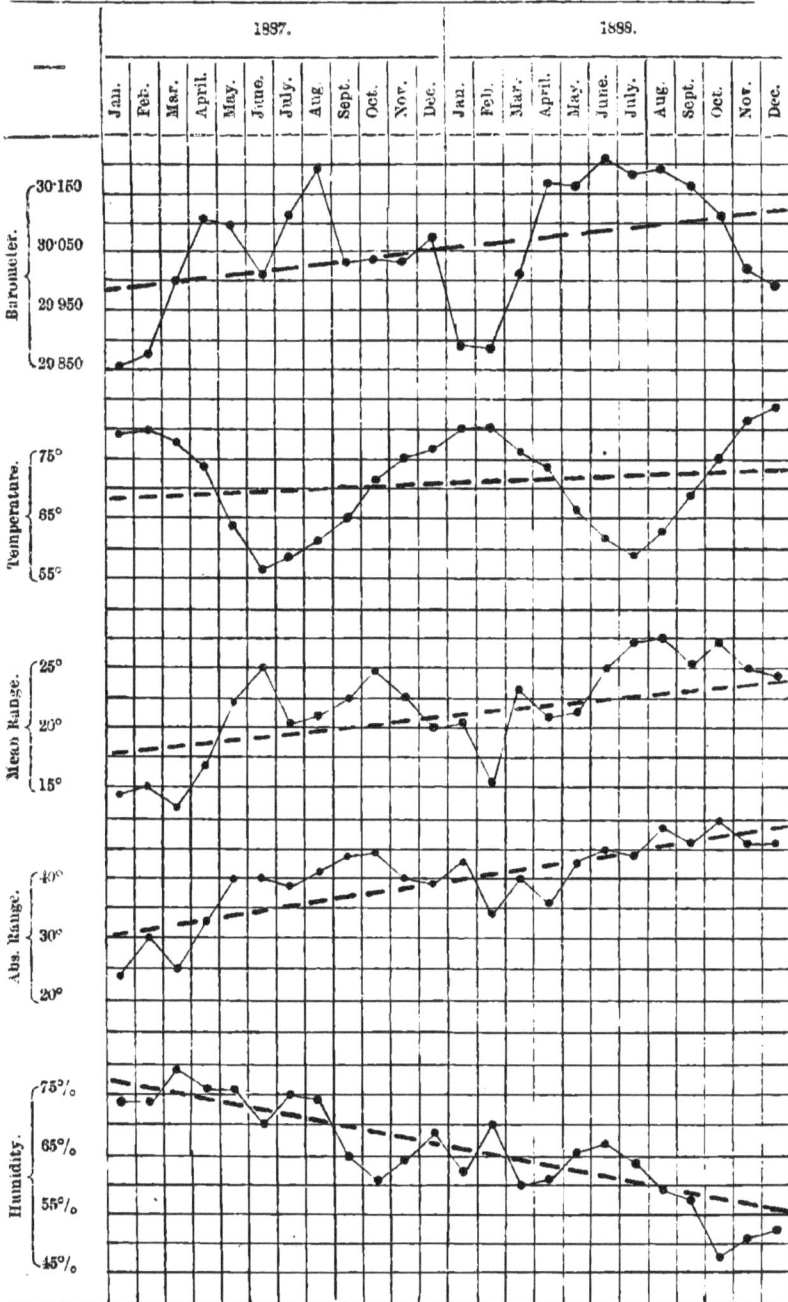

Typhoid Fever—Moreton Division (Brisbane).—The mean monthly curve of admission rate to hospitals in this district—Brisbane General Hospital and Ipswich Hospital—for the quinquennial period 1887-91 for typhoid fever shows a high maximum in January, February, and March, falls steadily from March to July, remains equally low August and September, and rises rapidly from September to December, the rate for the latter month being about equal to that of the following month of January.

The following tables will show the relation which the admission rate to hospitals bears to the total mortality according to population from one month to another, and for four periods of the year :—

MORTALITY and ADMISSION RATE PER 100,000 POPULATION PER ANNUM for EACH MONTH of the YEAR.

—	Jan.	Feb.	Mar.	April.	May.	June.	July.	Aug.	Sept.	Oct.	Nov.	Dec.
Mortality	95·4	111·6	95·7	59·8	42·0	42·5	34·6	24·3	21·5	21·5	52·4	87·2
Admissions ...	600·5	618·0	648·5	510·3	348·5	276·4	156·2	132·4	156·8	336·7	456·4	612·3

TABLE showing PROPORTION of TOTAL DEATHS to TOTAL NUMBER of ADMISSIONS.[1]

—	Jan.	Feb.	Mar.	April.	May.	June.	July.	Aug.	Sept.	Oct.	Nov.	Dec.
Deaths ...	10	10	10	10	10	10	10	10	10	10	10	10
Admissions ...	63	58	69	93	83	67	44	54	54	150	86	70

RATIO PER CENT. for FOUR PERIODS of the YEAR.

—	Jan. to March.	April to June.	July to Sept.	Oct. to Dec.	Total.
Deaths	43·9	20·9	11·8	23·4	100
Admissions ...	38·6	23·7	9·1	28·6	100

From these figures, as well as from an examination of the two curves, it will be found that there is a well-marked resemblance between the total number of admissions to hospital and the total mortality for the entire district all the year round. It may be observed, however, (1) that the month of highest mortality is February, while the number of admissions is equally high in February and March; (2) that the month of lowest death-rate is August, the admission rate being low in July, August, and September; and (3) that while the mortality remains low in October the admission rate takes a very decided rise during this month.

When the year is divided into the usual four periods it will be found that 43·9 per cent. of the total deaths for twelve months take place during the first, and only 11·8 per cent. during the third quarter of the year—that is to say, for every 10 deaths that occur the latter period there are *nearly* 40 the former. The admissions also form 38·6 and 9·1 per cent. respectively, or in other words for every 10 admissions in the months of July, August, and September there are a little *over* 40 from January to March. In comparing the

---

[1] *See* Footnote, p. 97.

mortality with the admission rate for the same periods it will be noticed that for every 63 admissions to hospital in the first quarter there are 10 deaths recorded by Registrar-General as occurring in the entire district, whereas the proportion is 51 to 10 in the third quarter or colder months of the year. This would appear to show, *if the same proportion of the cases that occur in general practice are regularly sent to hospital,* that the percentage of recoveries is greater in summer than in winter, but in referring to hospital statistics alone we shall find that this is not the case, and that on an average of five years the percentage of deaths is greater in the warm months of the year, as will be seen from the following figures :—

PERCENTAGE of DEATHS from TYPHOID FEVER and FEBRICULA (1887-91) occurring in HOSPITAL for FOUR PERIODS of the YEAR.

January to March.	April to June.	July to September.	October to December.
7·1	6·2	5·0	5·5

It follows, therefore, that not only an absolutely but relatively greater proportion of cases are sent to hospital in the summer months, and that when they occur in winter, medical men retain a larger number under their own charge in private practice.

The following tables will show the difference in both mortality and admissions from year to year, together with the meteorological observations for purposes of comparison :—

RATE PER 100,000 POPULATION for the YEARS as a Whole.

—	1887.	1888.	1889.	1890.	1891.
Admissions	384	372	756	312	204
Mortality	72	63	92	36	31

MEAN MONTHLY METEOROLOGICAL OBSERVATIONS for the YEARS as a Whole.

—	1887.	1888.	1889.	1890.	1891.
Barometer	30·036	30·128	30·077	30·039	30·071
Temperature	66·7°	68·6°	69·6°	68·7°	68·4°
Mean range	16·8°	19·1°	17·4°	17·3°	17·6°
Absolute range	34·1°	36·2°	35·0°	32·5°	32·7°
Humidity	72°/₀	66°/₀	69°/₀	72°/₀	69°/₀
Rainfall	7·150	2·960	4·115	5·160	4·030
Rainy days	20	13	12	14	12
Cloud	5·0	4·5	5·5	5·7	5·1

According to these statistics and observations, the year 1889 shows not only the greatest mortality but greatest number of admissions. From a meteorological point of view there is, with the exception of a high temperature, nothing to account in any way for the high rate of that year, for although the barometer and absolute range of temperature are fairly high they are higher still in 1888, while at the same time the humidity and rainfall during the latter year are particularly low. All these point to a high mortality, and when the mean monthly curves from one year to another are examined, it will at least be suspected that if these conditions do not directly affect the death-rate in the winter months of 1888—possibly because it is winter—they indirectly contribute towards that of the following year, 1889.

We shall find here that in connection with this period of high admission and death-rate both begin to rise in November, 1888; and in the case of the former the curve continues to rise to a high maximum in January, remains equally high till the end of April, and falls rapidly in May. The mortality curve, on the other hand, falls somewhat in the month of January, and remains at a high maximum only for the two months February and March. A fair idea of the magnitude especially of the admission curve will be formed by comparing it to those of other years. During the months of January, February, March, and April, 1889, the average number of monthly *admissions* to hospitals is 116, while that of other years is only 44, or, according to population, at the rate of 1,176 and 444 respectively per 100,000 per annum, the average number of admissions the former period being thus nearly *three* times greater than in the latter. At the same time the average monthly number of deaths is 15, while that of other years is 7, or at the rate of 156 and 74 respectively per 100,000 population per annum, there being thus for the first four months of 1889 fully *double* the average number of deaths in other years for the same period :—

—	January-April, 1889.	January-April, Other Years.
Monthly number of admissions	116	44
Monthly number of deaths [1]	15	7

Moreover, if the number of deaths be compared to the number of admissions, it will be found in the former case that there are 10 deaths for every 77 admissions, and that for other years the proportion is 10 to 60 respectively. This would appear to prove that a summer of high admission rate gives a lower percentage of deaths than one in which the disease is less prevalent, just as it at first sight appeared that the proportion of recoveries in summer is greater than in winter, and thus *apparently* showing that the period of highest rate generally, whether it be summer or winter, is the one attended by lowest mortality. We have found, however, in the latter instance that such is not the case—the percentage of deaths in summer being somewhat greater than in winter—and in referring to hospital statistics again, we shall find the same applies in comparing a summer of high admission rate with one of low admission rate. As an instance it may be mentioned that during the first quarter of 1889, when typhoid fever is very prevalent, the percentage of deaths in hospital from enteric fever (including febricula) is 6·7,

---

. [1] This is not hospital mortality alone, but includes also those deaths that occur in private practice as recorded by the Registrar-General for the entire district.

whereas in 1890 and 1891, for the same period, when the admission rate is low, it is only 3·6. It should be mentioned here that in the first quarter of 1887, when typhoid fever is not nearly so prevalent as in 1889, the hospital mortality is 13·3 per cent. It must be remembered, however, that the treatment by cold baths was not then applied to the same extent as at the present time, and if the death-rate is reduced by this means, we have here abundant testimony in favour of its extended use.[1]

In connection with this season of high mortality, the following are the meteorological observations for the first four months of 1889, and for comparison those also giving the mean of other years :—

MEAN MONTHLY METEOROLOGICAL OBSERVATIONS—JANUARY, FEBRUARY, MARCH, and APRIL.

					1889.		Mean of Other Years.
Barometer	...	...	...	...	30·075	...	30·015
Temperature	...	...	...	...	75·5°	...	73·6°
Mean range	...	...	...	...	17·1°	...	14·8°
Absolute range	...	...	...	...	32·0°	...	29·1°
Humidity	...	...	...	...	66%	...	74%
Rainfall	...	...	...	...	3·404	...	7·979
Rainy days	...	...	...	...	13	...	21
Cloud	...	...	...	...	5·9	...	6·6
Wind	...	...	...	...	9	...	11

This will show very fairly that during the months of high admission rate the barometric pressure and temperature, together with the mean daily and absolute range of temperature, are high, while the relative humidity, rainfall, number of rainy days, and amount of cloud are low. In looking carefully at the various meteorological curves, it will also be observed that for many months beforehand, perhaps even from the middle of 1887, the barometer and temperature have been gradually rising higher and higher—that is to say, the temperature in the winter months, 1888, is greater than that of 1887, and in the summer months, 1889, greater than that of 1888, and so also with the barometer—and the relative humidity and rainfall steadily falling, thus gradually tending in every respect, and to an extreme degree, towards that condition of the atmosphere that is associated with a high admission-rate from typhoid fever, and that in the absence of the usual fall of rain in summer, and under an exceptionally high temperature this rate rises to an extent at least three times greater than that observed during any other period for the last five years. (See p. 92.)

---

[1] In connection with this it may be mentioned that while the hospital mortality in the third quarter of the year is usually very low, there being no deaths recorded in 1887, 1890, and 1891—the admission rate being, however, also low, and the percentage of deaths for five years being on an average only 5—there were in 1889 out of 76 cases during these months 6 deaths, or at the rate of 7·9 per cent. If there is any connection between this increased mortality for that year during the winter and cold baths, it would follow that while highly beneficial in summer they are not attended by the same happy results when carried out in the cold months of the year. This is a matter to be determined only by those immediately in attendance upon the cases, and regarding which it would be well to receive more direct evidence than is possible from mere statistics.

In a carefully prepared paper by Dr. Hare ("Australasian Medical Gazette," July, 1889), read before the Queensland Medical Society, he pointed out that the hospital mortality of typhoid fever in Brisbane was reduced from an average of 14·8 per cent. in 1882-86 to 8·6 in 1887-88, evidently by means of cold bathing. He does not, however, enter into the question of seasonal death-rate.

With the exception of one month—August—the admission rate continues unusually high during the whole of 1889, indeed, shows an indication of rising higher than ever the following summer, and only falls in January, 1890, in sympathy with a corresponding fall in temperature, and abundance of rainfall, during the early months of the year.

The following table gives the ratio of north, north-east, and east winds during the first four months of 1889, and that also of other years.

TABLE showing the PREVALENCE of NORTH, NORTH-EAST, and EAST WINDS.

		1889.		MEAN—OTHER YEARS.	
		Times Observed.	Per cent. of Whole.	Times Observed.	Per cent. of Whole.
N.	January	21	31	6	9
	February	15	25	4	5
	March	3	4	2	3
	April	2	3	1	1
N.E.	January	13	18	14	19
	February	14	23	8	12
	March	8	11	5	7
	April	6	9	1	2
E.	January	14	20	25	34
	February	16	27	18	28
	March	17	24	17	25
	April	13	20	10	15
N.—Mean for January-April		10	16	3	5
N.E.—Mean for January-April		10	13	7	10
E.—Mean for January-April		15	23	17	26

It will here be seen that the north winds are, on an average, much more frequently observed in 1889 than in other years in the proportion of fully 3 to 1, and that north-east winds also are somewhat more prevalent, while the east winds are less common; thus showing, if there is any connection between these and the mortality from typhoid, that the north wind is favourable and the east wind unfavourable towards its development.

Typhoid Fever—Rockhampton Division.—The mean monthly mortality curve of typhoid fever for the years 1887-91 for this district shows a gradual incline from January to its maximum for the year in March, followed by a continuous fall to its minimum for the year in August, and then a rise from August to December, the rate for the latter month being fully higher than that of the following January.

MORTALITY PER 100,000 POPULATION PER ANNUM.

Jan.	Feb.	Mar.	April.	May.	June.	July.	Aug.	Sept.	Oct.	Nov.	Dec.
63	90	108	85	43	43	13	7	24	24	24	71

If we take the months of December, January, February, and March, and those of June, July, August, and September, and compare the mortality of one period with that of the other, we shall find that there are at the rate of 10 deaths during the latter for every 38 the former; or, if we take only January, February, and March, together with July, August, and September, the difference is greater still, the proportion being 10 to 60 respectively.

The following table for comparison gives the admission rate to hospitals for typhoid (including malarial fever) :—

ADMISSION RATE PER 100,000 POPULATION PER ANNUM.

Jan.	Feb.	Mar.	April.	May.	June.	July.	Aug.	Sept.	Oct.	Nov.	Dec.
576	540	816	720	750	430	250	216	180	280	360	576

This will show that there are 10 deaths recorded by the Registrar-General for every 76 admissions in December, January, February, and March, and 10 deaths for 123 admissions during the corresponding winter and spring months.

RATE PER 100,000 POPULATION for the YEARS as a Whole.

—	1887.	1888.	1889.	1890.	1891.
Admissions	696	372	612	264	396
Deaths	38	26	106	27	39

MEAN MONTHLY METEOROLOGICAL OBSERVATIONS for the YEARS as a Whole.

—	1887.	1888.	1889.	1890.	1891.
Barometer	...	30·097	30·080	30·025	30·074
Temperature	71·2°	71·3°	72·4°	72·1°	71·6°
Mean range	13·9°	14·7°	13·9°	17·7°	18·9°
Absolute range	28·4°	29·4°	29·0°	34·4°	34·1°
Humidity	75°/₀	74°/₀	74%	71°/₀	67%
Rainfall ... '	3·833	3·883	2·986	6·443	4·256
Rainy days	10	6	10	13	10
Cloud	4·1	3·8	4·7	5·1	4·8

It will be seen from these figures, not only that the highest mortality occurs in 1889, but that it is three times that of any other year, or nearly equal to the total mortality of the other four years combined. From a meteorological point of view this corresponds to a high temperature and low rainfall, although, taken *for the year as a whole*, it cannot be said that there is any connection with the remaining elements of the weather.

When the curve showing this high death-rate is examined monthly, it will be found to begin in December, 1888, to rise steadily to its maximum in March, 1889, and then to fall continuously to August. Its magnitude will be apparent when it is said that for the six months from December, 1888, to May, 1889, the average monthly mortality is at the rate of 197 per 100,000 per annum, wnile that of other years for the same period is only 47, the proportion being thus fully 4 to 1, or a little more than the total mortality for the other four years combined. The following table will show the meteorological conditions that obtain from December, 1888, to May, 1889, together with the mean of these for other years during the same period :—

				December, 1888, to May, 1889.	.	December to May, Other Years.
Barometer	...	...	...	30·051	...	29·999
Temperature	...	...	...	77·0°	...	75·8°
Mean range	...	...	...	11·1°	...	12·9°
Absolute range	...	...	...	25·9°	...	26·6°
Humidity	...	...	...	74%	...	73%
Rainfall	...	...	...	3·625	...	7·542
Rainy days	...	...	...	11	...	14
Cloud	...	...	...	5·2	...	5·4

It will here be noticed that the atmospheric conditions chiefly associated with the period of high mortality are high barometric pressure and temperature, and low rainfall. Moreover, with the exception of a very high rainfall in February, the rainfall in 1888 is particularly low, while the barometer is unusually high, and it is probable that these conditions, in the absence of the usual fall of rain the following summer, may contribute towards the high mortality of 1889. It will also be noticed that in December of the latter year the mortality has a tendency to rise high again, and apparently falls concurrently with the very heavy rainfall of January, February, and March of the following year, 1890.

REMARKS ON TYPHOID FEVER.

1. That it is essentially a disease of summer, and attains its maximum mortality and admission rate to hospitals in the month of February.

2. That according to hospital statistics alone the percentage of deaths is, on an average of five years, somewhat greater in summer than in winter, and the mortality is all the greater in summer, the more prevalent the disease and the larger the number of admissions.

3. That although the period of highest barometric pressure—winter—corresponds fairly to the period of lowest mortality, the higher the barometer during any season of the year the greater will be the death-rate.

4. That not only does the period of highest temperature correspond to the period of highest mortality, but the higher the temperature during any season of the year the greater will be the death-rate.

5. That although the period of highest mean daily and absolute range of temperature coincides with the season of lowest death-rate, the higher the range during any particular time of the year the higher will be the mortality.

6. That the period of lowest relative humidity corresponds to the period of highest mortality. It is possible that the latter is, however, due chiefly to the low rainfall and high temperature attending that period, for if the temperature be lower and the rainfall moderate a higher relative humidity is coincident with a higher death-rate.

7. That although the period of low rainfall, low number of wet days, and low amount of cloud corresponds to the season of low mortality, the lower these are during the months of summer the higher will be the death-rate.

8. That a high rainfall during the months of summer not only diminishes the mortality for that period of the year, but tends to do so also for the succeeding months of winter, even though the rainfall be then comparatively low.

9. That a moderately high rainfall during the winter months is not associated with a low death-rate if the temperature be comparatively high, possibly because the rainfall at that time is not sufficiently high to act the part of scavenger.

10. That for the district of Brisbane the north winds are very prevalent, and the east winds less frequent than usual, during the summer of highest mortality.

11. It follows from these observations—

(*a*) That the mean temperature of the air is the only element of the weather that is consistently attended by high or low mortality ; that is to say, the period of highest temperature not only corresponds to the season of highest mortality, but the higher the temperature during any particular time of the year the greater will be the death-rate. The remaining elements of the atmosphere, if *high* in summer—such as the rainfall—during the period of highest mortality, are attended by a lower mortality, the *higher* these are during any particular time of the year ; and conversely, if these elements be *low* in summer—such as the barometer and range of temperature—during the period of highest mortality, the *lower* they are at any time the lower will be the death-rate. All these elements together, therefore, counteract the influence of a high temperature in summer and help to keep down the mortality. If they fail to attend a high temperature in summer, the death-rate will rise very high.

(*b*) That a high mortality during any period of the year is associated with high barometric pressure and temperature, and high mean daily and absolute range of temperature, together with low relative humidity, rainfall, number of wet days, and low amount of cloud.

(*c*) That if these conditions continue and become more and more pronounced for a period of eighteen months—the seasonal fluctuations being, of course, excepted—it may safely be predicted that in the absence of the usual rainfall the following summer the mortality will rise exceedingly high.

(*d*) That a low mortality is dependent upon atmospheric conditions the opposite of these.

(*e*) That as the mean temperature and rainfall form the basis of the condition of the weather, a low mortality is primarily dependent upon a low temperature and high rainfall.

(*f*) That in all cases during the summer months there is probably a tendency to the development of those bacilli that cause typhoid fever, but under a heavy rainfall their germination stage is interfered with, and they are at once washed away.

(*g*) That in order to prevent an epidemic of typhoid fever during a dry summer, we should imitate Nature as much as possible by having our streets regularly watered and cleaned daily and our drains *largely* flushed out for two or three hours at a time at least twice a week, and preferably, on account of its germicidal properties, with salt water. The former will prevent the dust from being blown by the wind on the roofs of our houses, and thence into the water tanks, while the latter will act the part of public scavenger.

Malaria appears to be endemic only in Northern Queensland. Even there the number of deaths recorded by the Registrar-General is comparatively small, certainly not large enough to form a reliable basis for purposes of comparison with the condition of the weather.[1] It is different, however, with regard to hospital admissions, a large proportion of the cases admitted in some hospitals having a malarial origin.[2] This alone will show that though a common form of disease, it is rarely fatal. Unfortunately malaria is of such a varied type that it is not easy to say to what extent it affects or is associated with many febrile complaints that bear perhaps another name, and this difficulty is increased when it is remembered that each district or hospital has its own definition for what is probably one and the same disease. The following terms are generally made use of :—Malarial fever, malaria, fever, ague, fever and ague, gulf fever, remittent fever, and intermittent fever ; and as it is doubtful to what extent colonial fever, slow continued fever, and typhoid fever may be associated or influenced by the malarial poison, I have, in order to avoid confusion, for the following districts, classified all these as one under the general name "Malaria."[3] Moreover, we know that malaria is intimately connected with the opening up of new country and the first approach of civilisation into any particular district, and as this may and does go on with intermissions in a young colony, the prevalence of this disease in Queensland follows a similar and somewhat erratic course, independently even of the conditions of the weather. In this way we have possibly an explanation of the uncertain and variable relation that it seems to bear, in different localities, to atmospheric conditions, and which prevents us arriving at any reliable conclusions regarding the matter.

Malaria—Boulia and Blackall Division.—The mean monthly curve of admission rate to hospitals for this district for the years 1888-91 presents a high maximum for the months of March, April, and May, a low minimum from July to October, and a medium rate

---

[1] The following shows the percentage to deaths from all causes and the death-rate per 1,000 population for Queensland as a whole for the years 1887-91 :—

Malarial Fevers { Remittent Fever .. 0·78 } 0·83 per cent.
{ Intermittent Fever ... 0·05 }

The average for twelve civil divisions of India is 13·91 per cent. (Davidson), or nearly 17 times greater than for this colony. The death-rate for Queensland is only 0·13 per 1,000 per annum. In the European Army of India it is 1·42 (or fully 10 times greater), and among the civil population of India it is very much greater.

[2] For the years 1887-91 malarial fevers, including continued fevers of all kinds, formed 43 per cent. of all cases admitted to the Cairns District Hospital.

[3] "Flood fever" is the latest acquisition to our nomenclature for febrile complaints.

for the remaining months of the year, thus showing that it is distinctly an autumnal disease for this large district, as will be seen from the following:—

TABLE showing MEAN ADMISSION RATE PER 100,000 POPULATION PER ANNUM for EACH MONTH of the YEAR.

Jan.	Feb.	March.	April.	May.	June.	July.	Aug.	Sept.	Oct.	Nov.	Dec.
1,224·3	1,006·2	1,890·7	2,665·6	1,560·1	1,000·5	556·4	556·4	333·0	556·9	1,104·5	1.224·0

TABLE showing ADMISSION RATE PER 100,000 POPULATION for the YEARS as a Whole.[1]

1887.	1888.	1889.	1890.	1891.
3,197·6	890·5	950·8	1,356·7	1,344·0

According to these figures the year of highest admission rate is 1887, and of lowest 1888. The meteorological observations associated therewith are as follow :—

MEAN MONTHLY METEOROLOGICAL OBSERVATIONS for the YEARS as a Whole.

—	1887.	1888.	1889.	1890.	1891.
Barometer		30·021	29·975	29·951	29·999
Temperature		74·5°	75·6°	70·8°	70·3°
Mean range		31·2°	29·5°	23·4°	25·1°
Absolute range		52·5°	56·2°	43·7°	43·9°
Relative humidity		47°/₀	52°/₀	57°/₀	60°/₀
Rainfall	2·152	1·341	1·151	2·599	2·503
Rainy days	5	3	4	5	5
Cloud		1·8	2·5	3·3	3·2

In the above table the observations for 1887 are not complete, but we may observe that the rainfall and number of rainy days are about double that of 1888, the year of lowest admission rate. In 1890 and 1891 the number of admissions is also comparatively high, and here we find not only a high rainfall and large number of rainy days, but high relative humidity and high amount of cloud, together with low temperature and low mean and absolute range of temperature.

In comparing the monthly admission rate from one year to another, we find that the curve shows a high maximum in March, April, and May, 1890, and in March and April, 1891. It is found also that the rate is even much higher still during the autumn months

---

[1] The average admission rate per 100,000 population per annum for the years 1887·91 is 1,547,

of 1887, but as the meteorological observations, with the exception of rainfall, are not registered during that year it is not represented in the admission curve.

MEAN MONTHLY NUMBER of ADMISSIONS—MARCH, APRIL, and MAY.

1887.	1888.	1889.	1890.	1891.
78	14	11	26	22

MEAN ADMISSION RATE—MARCH, APRIL, and MAY—PER 100,000 POPULATION PER ANNUM.

1887.	1888.	1889.	1890.	1891.
9,468	1,632	1,248	2,844	2,228 ·

As regards the exceedingly high admission rate for 1887, we have unfortunately, except in rainfall, no meteorological observations to guide us as to its probable connection with the weather. It may be observed, however, that in March the rainfall is exceptionally high —13·880 inches—while the mean for the other four years is only 2·094 ; and that the number of rainy days is also comparatively high—15 against 5 in other years. Leaving, however, the admission rate for 1887 out of consideration, we may note that it is high in March, April, and May, 1890, and low in 1889. The following are the comparative observations for this period from one year to another :—

MEAN MONTHLY METEOROLOGICAL OBSERVATIONS—MARCH, APRIL, and MAY.

—	1888.	1889.	1890.	1891.
Barometer	30·071	30·026	29·993	30·032
Temperature	75·3°	76·2°	71·8°	70·6°
Mean range	29·8°	24·8°	20·6°	20·8°
Absolute range	50·3°	53·2°	38·5°	39·6°
Humidity	50°/₀	58°/₀	66°/₀	68°/₀
Rainfall	·455	2·043	3·224	4·181
Rainy days	1	4	7	7
Cloud	1·6	3·2	4·1	4·3
Wind	20½	16	12	16

It will be found from the above that a period of high admission rate is associated with low barometric pressure, low temperature and low mean and absolute range of temperature, together with high relative humidity, rainfall, number of rainy days, and cloud, and possibly also with low velocity of wind. It may also be observed that these same conditions exist during the two months, January and February previous to the rise in the admission rate, that the rainfall especially is high in January in both 1890 and 1891, and that while the temperature falls 9·7 deg. from December, 1889, to January, 1890, and 4·4 deg. from December, 1890, to January, 1891, it falls only ·8 deg. in

January, 1889. It is possible, therefore, that a heavy rainfall, combined with a fall in temperature of several degrees in January, may predispose towards a high admission rate in the following months of autumn.

Malaria—Cook District.—The mean monthly curve of admissions to hospitals for malaria for this district for the quinquennial period 1887-91 begins in January comparatively high, rises somewhat in February, and still higher to its maximum for the year in March, falls fairly regularly to its minimum for the year in August and September, and then rises from September to December, the rate for the latter month being a little below that of the following month of January. It is thus for this district both a summer and an autumnal disease.

The following tables based on the hospital returns for five years, give the admission rate according to population for each month of the year, and from one year to another respectively :—

MEAN ADMISSION RATE PER 100,000 POPULATION PER ANNUM.

Jan.	Feb.	Mar.	April.	May.	June.	July.	Aug.	Sept.	Oct.	Nov.	Dec.
5,637·6	6,697·2	8,400 ·0	5,637·6	7,017·6	5,209·2	3,721·2	2,337·6	1,919·6	2,550·0	2,061·6	4,146·0

MEAN ADMISSION RATE PER 100,000 POPULATION for the YEARS as a Whole.[1]

1887.	1888.	1889.	1890.	1891.
5,909·4	6,432·5	7,149·6	2,782·0	1,311·8

The latter shows that the admission rate is high during the first three and low during the latter two years. The meteorological observations connected therewith are :—

MEAN MONTHLY METEOROLOGICAL OBSERVATIONS for the YEARS as a Whole.

—	1887.	1888.	1889.	1890.	1891.
Temperature	77·9°	77·9°	79·6°	79·0°	77·1°
Mean range	13·3°	11·9°	11·2°	10·5°	9·7°
Absolute range	23·2°	24·5°	22·0°	22·3°	21·5°
Humidity	73%	72°/₀	73°/₀	68°/₀	68°/₀
Rainfall ...	5·526	4·981	4·366	6·167	5·758
Rainy days	11	10	9	10	11
Cloud	4·2	3·4	3·8	4·7	5·1

According to the above figures, there is really nothing except a somewhat higher relative humidity to consistently accompany a year of high or low admission rate, for although the highest temperature (79·6 deg.) is recorded during the year of highest number of admissions (1889), and the lowest temperature (77·1 deg.) during the year of

---

[1] The average admission rate per 100,000 population per annum for the district of Cook for the years 1887-91 is 4,716.

lowest admissions (1891), we find the temperature comparatively high in 1890, the admission rate being at the same time low. The remaining atmospheric conditions seem to be equally inconstant. It may at any rate be noted that these do not in any particular, except that of relative humidity, correspond with those observed for malaria in the district embraced by the two meteorological stations, Boulia and Blackall. In the latter case we have found that a year of high admission rate is associated with low temperature and low range of temperature, high relative humidity, rainfall, number of rainy days, and amount of cloud; in the above case we find it associated with a high relative humidity, but the rainfall is not high, and the temperature instead of being low is comparatively high. This, it is well to bear in mind, however, applies only to the years as a whole, and cannot therefore form a reliable test of the connection between malaria and the condition of the atmosphere.

In comparing the monthly admission rate from one year to another, it will be observed that the curve rises particularly high in the first few months of 1888, attaining its maximum in March; the number of monthly admissions from December, 1887, to the following May being 68, while that of other years during the same time is only 24, or at the rate of 7,500 and 2,544 per 100,000 population per annum respectively. The following table will give the meteorological observations for December, January, and February (summer), and for March, April, and May (autumn), during the period of high mortality, 1888, and also for comparison those showing the mean of other years :—

—	Summer, 1888.	Summer, Other Years.	Autumn, 1888.	Autumn, Other Years.
Temperature	81·8°	82·2°	79·1°	79·6°
Mean range	13·3°	12·4°	11·4°	10·3°
Absolute range	26·0°	22·4°	23·0°	19·6°
Humidity	80°/₀	71°/₀	74°/₀	75°/₀
Rainfall	13·120	9·882	5·723	9·065
Rainy days	18	13	15	18
Cloud	4·2	5·2	3·9	5·0

It will be observed from the above that in the months of summer, 1888, the temperature is somewhat lower, while the relative humidity, rainfall, and number of rainy days are decidedly higher than the mean of the other years combined, and in these respects they agree with what was observed during the period of high admission rate for the districts of Boulia and Blackall. In the present instance, however, there is no corresponding rise in amount of cloud, and the mean and absolute range of temperature, instead of being low, are somewhat above the average. If, again, we compare the admission rate in the autumn months with the meteorological observations we shall find that when at its height in 1888 the rainfall and number of rainy days are low. We may note, however, (1) that the rainfall of December, 1887, is especially high, being 16·870 inches, while the average for other years is only 5·367 inches; and (2) that it is preceded by seven months of very low rainfall, ·867 inch, while the average per month for other years is 1·660 inches. This would appear to show that an increase in the number of admissions is not associated with a high temperature, and that although the relative humidity, rainfall, and number of rainy days may or may not be high during the period of its highest intensity, the rainfall is unusually high two or three months beforehand, the effect of this high rainfall being all the greater if preceded by a period of drought. It is possible that this would indicate a germination stage in nature of one or two months' duration for the development of the miasm, or an incubation stage in the body of the same duration for the development of the disease.

Malaria—Cloncurry and Hughenden.—The mean monthly curve of admissions to hospitals for malaria for the years 1888-1891 is high in April, May, and June, low in August, and medium during the remaining months of the year, showing thus that it is a late autumnal disease for this large district.

ADMISSION RATE PER 100,000 POPULATION PER ANNUM.

Jan.	Feb.	Mar.	April.	May.	June.	July.	Aug.	Sept.	Oct.	Nov.	Dec.
612	612	906	1,344	1,272	1,524	852	240	490	540	540	906

ADMISSION RATE PER 100,000 POPULATION PER ANNUM.[1]

1888.	1889.	1890.	1891.
1,284	1,080	564	840

These figures show a high admission rate for the first two and a low rate for the last two years. The following are the meteorological observations on record for the years as a whole:—

MEAN MONTHLY METEOROLOGICAL OBSERVATIONS for the YEARS as a Whole.

—	1888.	1889.	1890.	1891.
Temperature	76·5°	78·7°	75·5°	72·8°
Mean range	27·5°	26·2°	24·9°	24·9°
Absolute range	44·9°	48·6°	46·0°	42·7°
Humidity	49%	53%	59%	54%
Rainfall	1·861	1·336	2·168	3·474
Rainy days	3	4	6	5
Cloud	1·5	2·2	2·9	2·8

[1] The average admission rate per 100,000 population per annum for this district for the years 1888-91 is 942.

It will here be observed that for this district a year of high admission rate is coincident with high temperature together with low relative humidity, rainfall, number of rainy days, and low amount of cloud. The temperature is, however, not highest, nor the rainfall lowest, during the year of highest rate (1888), nor do the opposite conditions obtain in 1890, the year of lowest admission rate, showing that over the extended period of twelve months there is no constant connection between the two. It would appear that in a general manner the atmospheric conditions in their relation to malaria are mainly in accord with those of Cooktown for the years as a whole, although it may be noted that in the above case a low relative humidity is connected with high admission rate—a condition not observed in the districts of either Cook, Boulia or Blackall.

An examination of the chart showing the monthly curve from one year to another will, however, supply us with somewhat more reliable information. Here it will be observed that in the months of March and April, 1888, and April, May, and June, 1891, the curve presents a high maximum, while in 1889 and 1890 it remains fairly low all the year. If we take the mean number of admissions for March, April, May, and June, 1888 and 1891 combined, together with those during the same months 1889 and 1890, we shall find that whereas in the former case the average number is 7 per month, in the latter it is only 3, or at the rate of 1,716 and 732 per 100,000 population per annum, respectively. The following tables will show the meteorological observations for 1888 and 1891 separately, and 1889 and 1890 combined, for both the summer and autumn months :—

SUMMER MONTHS.

—	1888.	1891.	1889-90.
Temperature	86·7°	82·8°	86·4°
Mean range	25·5°	21·1°	25·3°
Absolute range	39·6°	37·9°	44·3°
Humidity	56%	58%	53%
Rainfall	6·807	8·175	3·776
Rainy days	11	12	9
Cloud	3·4	4·6	3·4

AUTUMN MONTHS.

—	1888.	1891.	1889-90.
Temperature	75·6°	72·9°	77·4°
Mean range	24·2°	20·9°	20·5°
Absolute range	41·3°	36·1°	41·6°
Humidity	56%	63%	65%
Rainfall	·105	3·963	1·913
Rainy days	1	7	5
Cloud	1·2	3·7	3·0

It would thus appear from the above figures, as far as the two latter columns are concerned, that the high admission rate for April, May, and June, 1891, corresponds to a low temperature, low absolute range of temperature, high rainfall, number of rainy days, and high amount of cloud, not only in the autumn but summer months; while the low rate for 1889 and 1890 corresponds to conditions pretty well the reverse of these. It will also be noticed that the mean range of temperature is low, and the relative humidity high in summer though not in autumn, 1891, thus showing possibly that a high admission rate in the latter season may be intimately connected with the condition of the atmosphere that obtains during the previous months of summer. As regards the high admission rate in March and April, 1888, it is possibly connected solely with the unusually high rainfall and large number of rainy days in February, for with these exceptions there is nothing to account in any way for the comparatively high rate for this period.

Malaria—Roma and Thargomindah.—The mean monthly curve of admissions to hospitals, for the years 1887-91, is high in April, May, and June, and low the remaining months of the year.

MEAN ADMISSION RATE PER 100,000 POPULATION PER ANNUM.

Jan.	Feb.	March.	April.	May.	June.	July.	Aug.	Sept.	Oct.	Nov.	Dec.
96	84	96	528	456	216	72	36	48	72	96	98

The following gives the admission rate from one year to another according to population :—

ADMISSION RATE PER 100,000 POPULATION for YEARS as a Whole.[1]

1887.	1888.	1889.	1890.	1891.
444	108	96	50	144

showing that the highest rate occurs in 1887 and the lowest in 1890.

MEAN MONTHLY METEOROLOGICAL OBSERVATIONS for the YEARS as a Whole.

—	1887.	1888.	1889.	1890.	1891.
Temperature	66·7°	67·9°	70·5°	68·1°	67·7°
Mean range	24·8°	28·2°	24·1°	22·4°	24·8°
Absolute range	46·9°	51·4°	49·4°	44·2°	44·6°
Humidity	67%	50%	61%	66%	63%
Rainfall	2·244	1·031	1·814	3·503	2·717
Rainy days	5	3	5	6	5
Cloud	2·9	2·4	4·0	3·9	3·8

[1] The average admission rate per 100,000 population per annum for the years 1887-91 is 168.

The only elements of the weather that appear here to be associated with the high admission rate for 1887 are a low temperature and high relative humidity, but on the other hand the former is not at its highest and the latter is almost equally high in 1890, the year of lowest admission rate.

In examining the monthly admission curve it will be found that the high rate for 1887 chiefly occurs in April, May, and June, and the following table will show the difference in meteorological readings for February, March, April, and May of that year, and of the other years combined, for the same period.

	February to May, 1887.	February to May, 1888-91.
Temperature	69·9°	71·5°
Mean range	22·4°	22·2°
Absolute range	44·0°	42·3°
Humidity	72°/₀	64·6°/₀
Rainfall	2·896	3·006
Rainy days	5	6
Cloud	3·1	4·0

This again shows that the high admission rate in April, May, and June, 1887, is associated with a low temperature and high relative humidity. At the same time the rainfall, number of rainy days, and amount of cloud are low, and there does not appear generally to be a sufficient difference in the meteorological observations of this year to account for the high rate of admissions as compared to that of the remaining four years. It is probably, therefore, associated with some predisposing cause that is not here apparent.

Malaria—Normanton District.—The mean monthly curve showing the admission rate to hospitals for the years 1888-91 is somewhat erratic in its course, presenting as it does two maximum curves, the first embracing the months of May, June, and July, and the second that of October, while for the remaining months of the year it is comparatively low. It is thus a disease chiefly of winter and spring for this district.

MEAN ADMISSION RATE PER 100,000 POPULATION PER ANNUM.

Jan.	Feb.	Mar.	April.	May.	June.	July.	Aug.	Sept.	Oct.	Nov.	Dec.
2,856	3,432	3,144	2,856	5,436	5,436	5,436	3,144	3,720	6,012	4,008	2,292

ADMISSION RATE PER 100,000 POPULATION for the YEARS as a Whole.[1]

1888.	1889.	1890.	1891.
7,680	4,560	2,688	2,196

[1] The average admission rate per 100,000 population per annum for the years 1888-91 is 4,281.

It will here be seen that the years of highest rate of admissions are 1888 and 1889, and of lowest 1890 and 1891. The following are the meteorological observations associated therewith :—

MEAN MONTHLY METEOROLOGICAL OBSERVATIONS for the YEARS as a Whole.

—	1888.	1889.	1890.	1891.
Barometer	30·025	29·993	29·935	29·989
Temperature	81·0°	83·1°	79·6°	77·9°
Humidity	65°/₀	69°/°	58°/₀	58°/₀
Rainfall	3·106	1·311	5·795	4·581
Rainy days	5	4	7	5
Cloud	2·0	2·7	3·5	3·9
Wind	14	14	12½	16

The relation here presented is very doubtful. It is true that if the atmospheric conditions of 1889 be compared to those of 1890 and 1891, it would appear that a high rate of admissions, is associated with high temperature and high relative humidity, together with low rainfall, number of rainy days, and low amount of cloud; but if compared on the other hand to those of 1888, where the rate is higher still, the temperature and relative humidity are not quite so high and the rainfall is slightly higher. For the years as a whole, therefore, there is no constant connection between the two.

In examining the monthly curve from one year to another, it will be noticed that the large number of admissions for 1888 is made up not so much by an unusually high rate during any particular season, but a generally high rate for the whole year. It may merely be noted in connection with this, that after a heavy rainfall in January and February there is practically no rainfall for the succeeding eight months—certainly a longer period of dry weather than observed any other year. Whether the large number of admissions in 1888 is connected with either or both of these factors it is not easy to say, but at least it should be observed, that in 1889 when the rate is also comparatively high the rainfall in summer is very low, while in 1890 and 1891, when the rate is very low, the rainfall in January and February is very high. In the last two years, however, there is not such a long period of dry weather in the winter months as noticed in 1888. The temperature during the years of highest admission rate is unusually high in summer, winter, and spring.

REMARKS ON MALARIA.

1. That for the colony of Queensland the meteorological observations in their relation to malaria are not a constant quantity.

It may be stated in a general way, however—

2. That malaria is chiefly an autumnal disease, the period of maximum rate of admissions being from March to June, and coincident with the period of lowest mean and absolute range of temperature, and highest relative humidity.

3. That a season of high admission rate is associated at the time with low barometric pressure, low temperature and low mean and absolute range of temperature, together with comparatively high relative humidity and rainfall, and large number of wet days.

4. That a season of high admission rate is preceded in summer by low temperature, high rainfall, and great number of rainy days.

5. That a low temperature, high rainfall, and great number of wet days in summer, do not tend to raise the rate of admissions so much at the time, as they do a month or two afterwards.

6. That the rate will possibly be all the higher if this high rainfall is preceded by a period of drought.

7. That a rise in the admission rate subsequent to high rainfall is not necessarily connected with the drying process, as the rainfall may—though not always—continue fairly high all through the period of its greatest intensity. Should the high rainfall be followed by a period of dry weather however, malaria will probably be all the more prevalent.

TOTAL AND COMPARATIVE MORTALITY.

Total Mortality—Queensland.—The mean monthly curve of total mortality for Queensland as a whole for the years 1887-91 shows a high maximum in January, February, and March, a gradual decline to its minimum for the year in September, followed by a sudden rise to November and December; the period of highest mortality being from November to March, and the months of lowest death-rate August and September. This, it may be observed, follows pretty nearly the same course as the curve of gastro-enteric diseases, showing that these are the most important diseases met with, and that their death-rate provides a fair index of the magnitude of the general mortality. This is not to be wondered at when it is remembered—as we shall see later on—that diarrhœal diseases alone form 12·63, and all gastro-intestinal diseases 22·77, or, including enteric fever, 25·64 per cent. of the total deaths from all causes. It is well to note here, however, that the mortality from other diseases—such as those of the nervous and urinary systems, and accidents of all kinds[1]—is higher during the summer months, and that these contribute somewhat to form the general mortality curve after the manner of that for gastro-intestinal diseases.

In the following page is a table which gives the total mortality from all diseases for Queensland as a whole, and for its various divisions for the years 1887-91 according to population.

---

[1] In this connection it may be noted here that, according to the Registrar-General's statistics, accidents—leaving out of consideration cases of sunstroke and drowning, which may naturally be expected to occur more frequently in early summer—are of more common occurrence during the fourth quarter of the year, as will be seen from the following :—

RATIO PER CENT. during FOUR PERIODS of the YEAR.

January to March.	April to June.	July to September.	Oct. to December.	Total.
23·0	23·8	24·8	28·4	100

This may be due directly to the condition of the nervous system, which we have seen is materially influenced by the state of the atmosphere at this season of the year.

TOTAL MORTALITY PER 1,000 OF MEAN POPULATION PER ANNUM FOR FIVE (5) YEARS—1887-1891.

District.	Group.	Jan.	Feb.	March.	April.	May.	June.	July.	Aug.	Sept.	Oct.	Nov.	Dec.	Mean for Year.
Cook, &c. ...	1	17·34	23·85	28·72	27·10	23·02	23·85	21·86	21·85	23·84	21·85	19·51	21·86	22·86
Burke ...	2	48·79	33·07	52·59	32·53	29·28	21·14	25·48	30·90	26·57	37·68	39·58	40·39	34·70
Cloncurry, &c.	3	14·63	22·49	18·78	11·57	19·87	15·07	19·87	11·57	13·54	18·78	12·45	24·02	17·03
Blackall, &c. ...	4	16·11	18·13	19·56	13·81	11·08	15·15	20·72	12·18	11·99	14·00	14·00	13·33	14·96
Balonne, &c. ...	5	13·04	13·04	12·17	9·92	14·05	10·58	12·53	11·30	9·20	7·68	7·46	13·18	11·15
Darling Downs, &c. ...	6	8·28	6·74	7·91	6·47	7·49	7·49	6·11	5·84	4·73	5·58	5·74	7·75	6·69
Townsville, &c.	7	20·00	20·13	20·00	18·51	16·61	16·40	15·47	15·12	19·86	18·51	15·59	21·29	18·15
Rockhampton, &c. ...	8	22·00	16·29	17·23	19·43	16·03	17·45	19·95	19·80	13·78	17·23	20·59	19·22	18·30
Brisbane, &c. ...	9	17·96	16·16	16·00	16·21	15·09	15·40	15·42	14·20	15·90	19·93	22·14	19·91	16·89
Queensland ...	10	16·48	15·44	16·62	15·34	14·61	14·12	14·61	13·63	13·14	14·75	16·38	16·26	15·11

These figures show :—

1. That for Queensland as a whole the death-rate varies from an average of 13·38 in August and September to an average of 16·24 from November to March per 1,000 population, and that for the twelve months the mean rate is 15·11.[1]

2. That for the Moreton District (Brisbane) the death-rate varies from a minimum of 14·20 in August to an average of 19·98 from October to January—the months of highest mortality—the mean for the year being 16·89.

3. That for the district of Rockhampton the rate varies from a minimum of 13·78 in September to an average of 20·90 from November to January, the mean for the year being 18·30.

4. That for the district of Townsville the mortality varies from an average of 15·29 in July and August to an average of 20·35 from December to March, the mean for the year being 18·15.

5. That for the Darling Downs the mortality varies from a minimum of only 4·73 in September to a maximum of 8·28 in January, the mean for the year being 6·69.

6. That for the district of Roma and Thargomindah the death-rate varies from an average of 7·57 in October and November to a maximum of 14 05 in May, the mean for the year being 11·15.

7. That for the district of Boulia and Blackall the mortality varies from a minimum of 11·03 in May to a maximum of 20·72 in July, the mean for the year being 14·96.

8. That for the district of Cloncurry and Hughenden the death-rate varies from a minimum of 11·57 in April and August to a maximum of 24·02 in December, the mean for the year being 17·03.

9. That for the district of Cook the death-rate varies from a minimum of 17·34 in January to an average of 27·91 in March and April, the mean for the year being 22·86.

10. That for the district of Burke (Normanton) the mortality varies from a minimum of 21·14 in June to a maximum of 52·59 in March, the mean for the year being as high as 34·70.

11. It will thus be seen that the general yearly mortality is lowest in West Southern Queensland (Darling Downs and Warrego), the average for these districts being 8·92 or a little over *half* the rate for the colony as a whole, and that it is highest in West Northern Queensland (Normanton), being 34·70, or fully *double* that of Queensland as a whole.

---

[1] The death-rate in the United Kingdom (1886-90), and in some of the Australian Colonies (1887-91), is as follows :—

England	... 18·9
Scotland	... 18·8
Ireland	... 17·9
Victoria	... 16·27
Queensland	... 15·11
New South Wales ...	... 13·45
South Australia ...	... 12·67

Comparative Mortality.—The following table will show the percentage of deaths from certain diseases to deaths from all causes for the colony as a whole for the years 1887-91; and to which may be annexed another (*see* p. 119), giving more in detail the percentage of deaths for each division of Queensland for each month of the year and for the year as a whole, based on an average of five years :—

AVERAGE PERCENTAGE to DEATHS from ALL CAUSES, QUEENSLAND, 1887-91.

Diphtheria	2·13
Diphtheria, croup, and laryngitis combined	3·64
Phthisis	8·75
Respiratory diseases	9·20
All respiratory diseases (including phthisis)	17·95
Typhoid fever	2·87
Diarrhœal diseases { Cholera nostras ... ... 0·52 / Diarrhœa ... ... 6·35 / Dysentery ... ... 3·09 / Enteritis ... ... 2·67 }	12·63
Other diseases of the digestive system	10·14
All gastro-intestinal diseases (including diarrhœal diseases) ...	22·77
All gastro-intestinal diseases (including diarrhœal diseases and enteric fever)	25·64

These figures, together with those in the following page, will show :—

1. That for the whole of Queensland, diarrhœal diseases stand highest in point of general mortality, forming as they do 12·63 per cent. of all deaths, or, including all gastro-enteric diseases 25·64 per cent. ; that respiratory diseases come next with a percentage of 9·20, and followed closely again by phthisis with 8·75, the rate for all respiratory diseases including phthisis, being thus 17·95.

2. That acute respiratory diseases (bronchitis, pneumonia, and pleurisy) are prevalent all over the colony, and vary from a minimum percentage rate of 6·80 in the district of Cook, to a maximum rate of 12·49 in the district of Rockhampton.

3. That phthisis is common along the eastern portion of the colony from Cooktown to Brisbane, averaging 11·62 per cent. of all deaths and attaining a maximum rate of 12·86 in the Rockhampton District.[1] That for the western portion of Queensland it is much less prevalent, forming as it does on an average only 2·99 per cent. of all deaths, and falling to the minimum of 1·56 in the district of Normanton.

4. That with the exception of the Darling Downs and Warrego Districts (West Southern Queensland), where the percentage is only 4·49, diarrhœa and dysentery are prevalent all over the colony, more especially along the eastern and northern seaboard (including Normanton), and attain the maximum rate of 11·75 in the Moreton District (Brisbane).

---

[1] The higher mortality for the districts of Rockhampton, Townsville, and Cooktown as compared to Brisbane is no doubt due to the larger Polynesian population on sugar plantations in these districts. It is well known that kanakas are particularly liable to contract phthisis, and statistics show that the death-rate amongst them from this cause is high.

PERCENTAGE TO DEATHS FROM ALL CAUSES—QUINQUENNIAL PERIOD, 1887-1891.

Disease	District	Group	Jan.	Feb.	March	April	May	June	July	Aug.	Sept.	Oct.	Nov.	Dec.	Mean for Year
Acute Respiratory Diseases	Cook, &c.	1	1·23	8·26	4·51	4·64	5·61	7·27	9·90	14·14	5·50	2·00	11·11	7·92	6·80
	Burke	2	2·67	7·84	9·88	2·04	13·33	6·03	5·71	14·58	12·20	6·00	6·56	4·84	7·51
	Cloncurry, &c.	3	4·28	2·33	2·78	18·18	10·53	6·90	7·80	22·73	11·11	11·11	12·50	13·04	10·29
	Blackall, &c.	4	1·43	5·00	5·58	6·67	12·50	13·64	8·89	13·21	19·23	13·11	4·92	12·07	0·20
	Balonne, &c.	5	6·07	5·33	12·86	14·04	12·35	20·34	10·44	13·21	13·31	20·45	9·30	7·89	11·53
	Darling Downs, &c.	6	1·54	7·65	11·19	11·76	20·34	20·34	14·68	17·39	8·11	6·82	8·39	6·58	11·27
	Townsville, &c.	7	0·60	4·03	3·25	8·77	2·04	8·91	12·63	10·75	11·99	8·77	9·28	4·69	7·16
	Rockhampton, &c.	8	7·43	9·91	5·98	7·17	10·39	19·33	22·70	21·13	10·39	10·21	11·75	8·40	12·10
	Brisbane, &c.	9	5·16	5·24	7·87	9·04	8·62	12·97	13·63	13·23	15·60	4·87	6·79	7·16	8·49
Phthisis	Cook, &c.	1	14·81	9·42	9·02	8·45	14·95	14·55	11·88	0·00	2·44	12·03	14·44	19·81	12·50
	Burke	2	2·67	1·08	2·47	2·04	0·00	3·03	0·00	0·00	3·70	1·72	1·64	0·00	1·50
	Cloncurry, &c.	3	0·00	2·33	2·78	0·00	2·63	0·00	5·28	3·96	0·00	5·56	0·00	2·17	2·83
	Blackall, &c.	4	1·43	1·27	2·36	5·00	6·25	3·03	9·33	4·62	5·66	1·64	2·33	3·46	2·55
	Balonne, &c.	5	0·00	1·33	1·61	1·75	6·17	4·92	4·28	4·31	16·22	0·00	8·89	6·59	3·89
	Darling Downs	6	1·54	3·77	7·11	3·92	3·39	5·06	2·08	15·05	18·03	2·27	7·22	1·64	4·13
	Townsville	7	9·70	13·71	11·39	7·02	9·90	7·02	17·89	15·96	15·96	17·54	12·91	12·98	12·40
	Rockhampton	8	10·81	10·36	12·34	10·94	12·99	10·92	11·40	10·61	11·17	13·62	8·42	16·03	12·86
	Brisbano	9	7·80	8·30	10·84	9·81	7·80	9·53	9·77	7·07	7·32	6·60	2·78	7·80	8·72
Diarrhoeal Diseases	Cook, &c.	1	8·64	3·67	9·77	9·63	7·19	10·91	5·94	6·25	7·10	14·00	3·28	7·92	9·34
	Burke	2	18·00	7·84	13·59	14·29	13·33	3·03	20·00	0·00	7·40	5·17	0·00	0·45	9·86
	Cloncurry, &c.	3	3·57	2·33	8·33	4·55	10·53	55·17	5·20	13·21	3·85	5·56	9·81	8·68	0·68
	Blackall, &c.	4	14·29	5·00	7·00	8·33	6·25	6·25	7·78	3·04	5·03	4·92	0·00	5·17	7·41
	Balonne, &c.	5	8·00	6·67	7·14	7·02	3·70	4·92	4·16	0·00	2·70	4·50	2·22	5·26	5·19
	Darling Downs, &c.	6	8·15	7·55	8·06	3·92	0·00	1·69	0·00	8·60	8·20	0·00	7·22	9·84	3·81
	Townsville	7	9·76	6·15	9·75	11·40	7·84	7·92	5·26	1·85	3·10	9·65	17·70	4·59	8·07
	Rockhampton	8	20·95	1·87	15·74	9·43	6·19	4·62	5·51	4·05	8·70	14·04	20·28	11·45	10·75
	Brisbane	9	11·81	11·24	9·94	12·19	0·17	6·43	5·49	0·00	0·00	20·35	1·11	13·76	11·75
Typhoid fever	Cook, &c.	1	0·00	1·63	1·20	1·61	0·00	0·90	0·00	0·00	2·44	2·00	0·00	0·99	0·67
	Burke	2	1·33	0·90	1·23	2·04	0·00	0·00	0·00	4·55	3·70	1·72	4·17	0·45	1·41
	Cloncurry, &c.	3	0·00	0·00	0·03	0·00	5·26	6·99	0·00	0·01	0·00	0·00	3·29	0·00	1·80
	Blackall, &c.	4	4·29	0·00	5·89	1·07	6·25	4·55	3·33	1·54	1·89	1·64	2·33	2·63	2·81
	Balonne, &c.	5	1·33	1·33	0·00	0·00	0·00	3·28	1·39	1·08	0·00	0·00	2·22	0·00	1·30
	Darling Downs, &c.	6	1·54	9·13	0·00	0·00	2·94	0·00	0·00	0·37	0·82	0·00	4·12	0·76	1·11
	Townsville, &c.	7	4·88	2·12	1·63	1·75	2·60	2·97	3·16	1·74	1·50	0·88	1·07	3·92	2·24
	Rockhampton, &c.	8	3·04	5·86	6·38	4·53	2·87	2·52	0·74	1·91	1·83	1·28	2·43	4·39	2·77
	Brisbane, &c.	9	6·39	7·02	6·06	3·69	2·90	2·83	2·29	1·01	0·00	1·30	0·00	1·61	3·47
Diphtheria	Cook, &c.	1	1·23	0·92	0·00	0·0	0·00	2·73	0·00	0·00	0·00	0·00	0·00	0·00	0·71
	Burke	2	0·00	5·88	2·17	2·04	0·00	0·00	0·00	0·00	0·00	0·00	0·00	0·00	1·10
	Cloncurry, &c.	3	0·00	2·33	0·00	0·00	0·00	0·00	0·00	6·16	0·00	0·00	0·00	0·00	0·26
	Blackall, &c.	4	0·00	0·00	0·00	0·00	0·01	1·64	1·11	2·17	3·85	1·64	0·00	2·03	0·51
	Balonne, &c.	5	0·00	0·00	0·00	5·20	8·73	6·73	2·78	0·00	0·00	4·50	0·00	1·64	1·81
	Darling Downs, &c.	6	6·15	0·00	3·23	0·00	4·90	0·00	6·25	6·16	0·00	2·27	2·40	5·31	3·17
	Townsville, &c.	7	0·00	0·00	0·61	0·88	2·60	1·26	1·05	2·17	3·28	0·00	1·79	1·15	1·67
	Rockhampton, &c.	8	0·00	1·35	0·85	1·51	4·65	6·17	1·81	1·85	1·04	0·85	0·74	1·08	1·34
	Brisbane, &c.	9	1·38	1·40	2·45	3·19			4·60	3·49	2·34	1·19			2·59

5. That deaths from typhoid fever form only 2·87 per cent. of the total mortality for the whole of Queensland, and vary from a minimum rate of 0·87 for the district of Cook to a maximum rate of 3·47 for the district of Moreton (Brisbane).

6. That for the whole of Queensland, the number of deaths from diphtheria, croup, and laryngitis combined is in excess of the deaths from typhoid fever, forming 3·64 per cent. of the total mortality, the maximum rate according to population being observed in the Darling Downs and Moreton Districts.

GENERAL SUMMARY.

1. The average total mortality for Queensland as a whole (1887-91) is 15·11 per 1,000 population per annum; the months of highest death-rate being November, December, January. February, and March, and of lowest August and September.

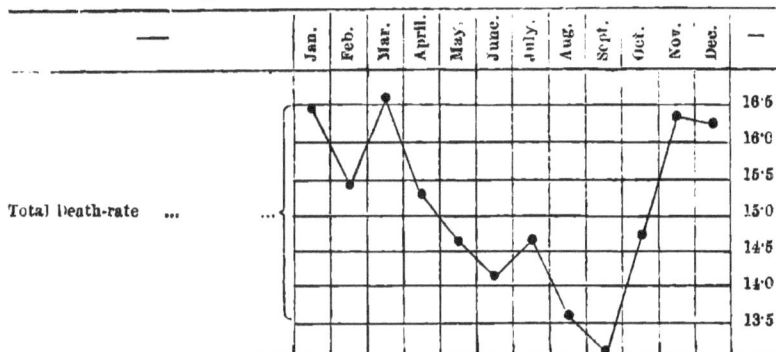

—	Jan.	Feb.	Mar.	April	May.	June.	July.	Aug.	Sept.	Oct.	Nov.	Dec.	—
Total Death-rate													16·5 16·0 15·5 15·0 14·5 14·0 13·5

2. The mortality from gastro-enteric diseases (1887-91) for Queensland as a whole is at the rate of 3·86 per 1,000 population per annum, forms 25·64 percentage to deaths from all causes, and attains its maximum during the hot period of the year; the months of highest death-rate from enteric fever being December, January, February, and March, from diarrhœa, October, November, December, and January, and from dysentery, November, January, and February.

DEATH-RATE from GASTRO-ENTERIC DISEASES—1887-91.

	Per 1,000 Population per Annum.	Percentage to Total Deaths.
Diarrhœal diseases	1·90	12·63
Other diseases of the digestive system ...	1·53	10·14
Enteric fever	·43	2·87
Total	3·86 ...	25·64

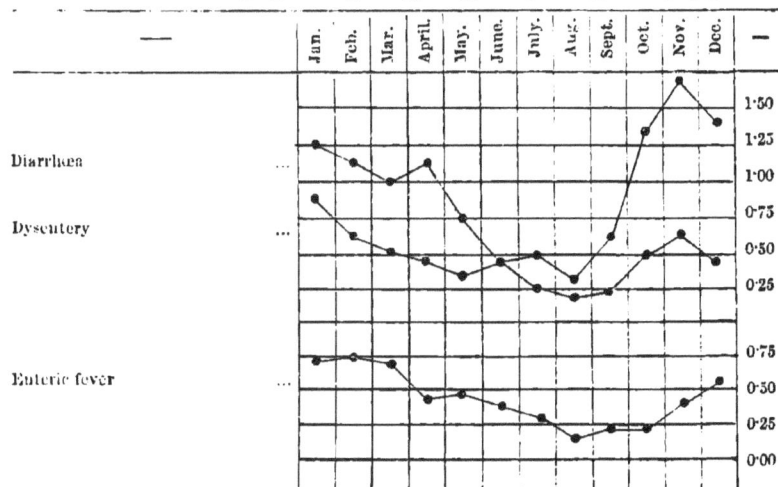

—	Jan.	Feb.	Mar.	April	May.	June.	July.	Aug.	Sept.	Oct.	Nov.	Dec.	—
Diarrhœa ...													1·50 1·25 1·00
Dysentery ...													0·75 0·50 0·25
Enteric fever ...													0·75 0·50 0·25 0·00

K

3. The mortality from all respiratory diseases (1887-91) for Queensland as a whole is at the rate of 2·71 per 1,000 population per annum, forms 17·95 percentage to deaths from all causes, and attains its maximum intensity during and immediately after the colder period of the year, the months of highest death-rate from pneumonia being July and August; from bronchtis, June, July, August, and September; and from phthisis, July, August, and September.

DEATH-RATE FROM ALL RESPIRATORY DISEASES (1887-91).

	Per 1,000 Population per Annum.	Percentage to Total Deaths
Phthisis [1]	1·33	8·75
Pneumonia	·54	3·82
Bronchitis	·46	3·04
Other respiratory diseases	·38	2·34
Total	2·71	17·95

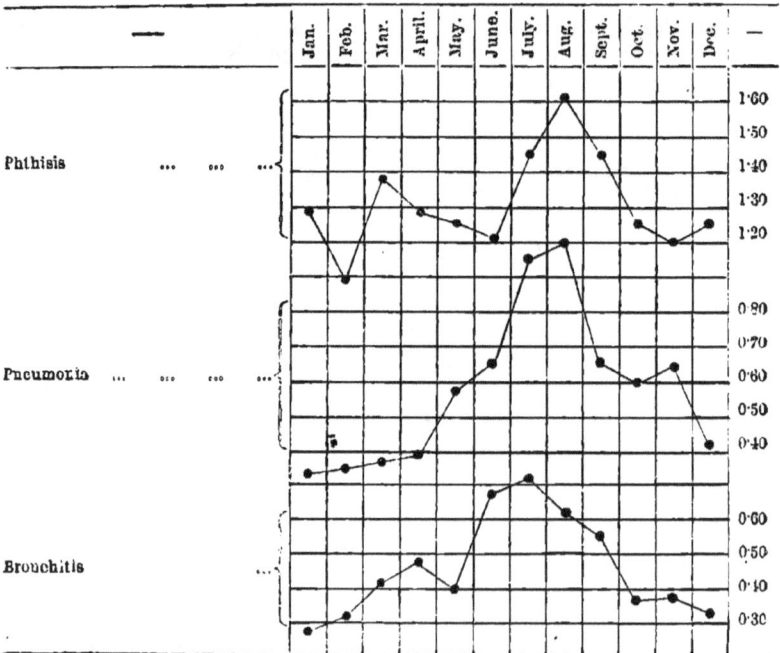

[1] Although not properly a disease of the respiratory organs in the same sense as pneumonia and bronchitis are, I have for obvious reasons included phthisis in this table. For similar reasons enteric fever is included in the table and tracing of gastro-enteric diseases. (*See* p. 121.)

4. The mortality from whooping-cough (1887-91) for Queensland as a whole is at the rate of ·14 per 1,000 population per annum, forms 0·93 percentage to deaths from all causes, and rises to its maximum, not (as observed for respiratory diseases generally) during the cold but warm period of the year; the months of highest mortality being December, January, and February.

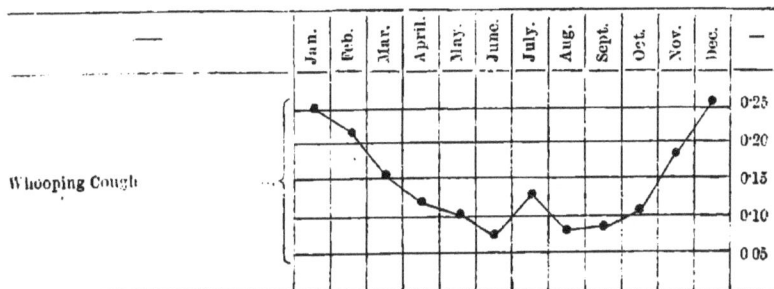

5. The mortality from diphtheria, croup, and laryngitis together (1887-91) for Queensland as a whole is at the rate of ·55 per 1,000 population per annum, forms 3·64 percentage to deaths from all causes, and attains its maximum during the colder period of the year; the months of highest death-rate from diphtheria being May, June, and July, and from croup, April, May, June, and July.

DEATH-RATE from ACUTE LARYNGEAL DISEASES (1887-91).

	Per 1.000 Population per Annum.		Percentage to Total Deaths.
Diphtheria	·32	...	2·14
Croup	·18	...	1·18
Laryngitis	·05	...	·32
Total	·55	...	3·64

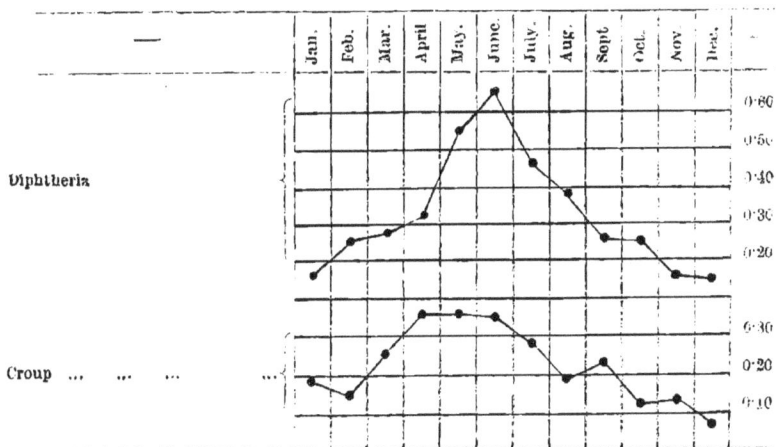

Mortality in relation to Atmospheric Conditions—In the preceding pages a period of high mortality from certain diseases is fixed upon, the meteorological observations in connection with these obtained, and, in comparing the one with the other, certain conclusions arrived at. These may here be given briefly in a tabulated form :—

Meteorological Observations.	Typhoid Fever.	Diarrhœal Diseases.[1]	Respiratory Diseases.	Diphtheria.	Whooping Cough.
Barometer	high ...	high ...	high ...	low ...	low
Temperature	,, ...	,, ...	low [2] ...	low [3] ...	medium [4]
Mean range	,, ...	,, ...	high ...	,, ...	low
Absolute range	,, ...	,, ...	,, ...	,, ...	,,
Humidity	low ...	low ...	low ...	high ...	high
Rainfall	,, ...	,, ...	,, ...	,, ...	,,
Rainy days	,, ...	,, ...	,, ...	,, ...	,,
Cloud	,, ...	,, ...	,, ...	,, ...	,,

Should these conclusions be even approximately correct, it follows by reversing the method of comparison, and having supplied to us, as a basis to start upon, certain meteorological observations, that it is possible to form some practical idea of the diseases that may be expected or associated therewith.

It will here be found —

1. That under the following atmospheric conditions in summer, namely :—

> High barometric pressure,
>   ,,   mean temperature,
>   ,,   mean daily range of temperature,
>   ,,   absolute range of temperature,
> Low relative humidity,
>   ,,   rainfall,
>   ,,   number of rainy days,
>   ,,   amount of cloud,—

we may expect a high mortality from typhoid fever, diarrhœa, and dysentery, as well as a somewhat higher death-rate from pneumonia and phthisis, while that from diphtheria will be unusually low.

2. That should the above conditions obtain also during the previous months of spring, the mortality in summer from typhoid fever, diarrhœa, and dysentery will be all the higher, while that from diphtheria will be similarly reduced.

---

[1] The opposite conditions may be associated with a high mortality in spring and early summer.

[2] Pneumonia may be connected with a high temperature in summer.

[3] If the temperature is below a certain point (58°) in winter the mortality falls.

[4] Low in summer and high in winter.

3. That under the following atmospheric conditions in summer, namely :—

> Low barometric pressure,
>   „ mean temperature,
>   „ mean daily range of temperature,
>   „ absolute range of temperature,
> High relative humidity,
>   „ rainfall,
>   „ number of rainy days,
>   „ amount of cloud,—

the mortality from typhoid fever, diarrhœa, and dysentery is low, while that from diphtheria and whooping-cough is equally high.

4. That should these conditions prevail during the former months of spring, the death-rate in summer from typhoid, diarrhœa, and dysentery will be still further reduced, while that from diphtheria and whooping-cough will be still further increased.

5. That should these conditions be decidedly well marked in summer—as in a time of flood—the mortality from diphtheria will probably be high during the months of autumn.

6. That the condition of the atmosphere, therefore, during the last few months of the year, provides a fair means of estimating the class of disease that may be expected during the early months of the following year, and similarly also its condition in summer provides a fair means of estimating the class of disease likely to be prevalent the following autumn.

7. That under the following atmospheric conditions in autumn and winter, namely :—

> High barometric pressure,
> Low mean temperature,
> High mean range of temperature,
>   „ absolute range of temperature,
> Low relative humidity,
>   „ rainfall,
>   „ number of rainy days,
>   „ amount of cloud,—

we may expect an increased mortality from respiratory diseases, together with a low death-rate from diphtheria, croup, and whooping-cough.

9. That under the following atmospheric conditions in winter, namely :—

> Low　barometric pressure,
> „　mean temperature,
> „　mean range of temperature,
> „　absolute range of temperature,
> High relative humidity,
> „　rainfall,
> „　number of rainy days,
> „　amount of cloud,—

the mortality from respiratory diseases will be reduced, and that from diptheria will be increased.

10. That under the following atmospheric conditions in winter, namely :—

> Low barometric pressure,
> High mean temperature,
> Low mean range of temperature,
> „　absolute range of temperature,
> High relative humidity,
> „　rainfall,
> „　number of rainy days,
> „　amount of cloud,—

the mortality from respiratory diseases will be reduced still further, while whooping-cough, on the other hand, will possibly become prevalent during the following months of spring.

APPENDIX.

THE FLOOD OF 1893.

(Read before the Queensland Medical Society, 14th March, 1893.)

DURING the first and third week of February last, Brisbane was visited by the most disastrous floods ever experienced by Europeans here, whereby many of the low-lying parts of the town and suburbs were submerged in water, and subsequently, as the water receded, covered with foul-smelling mud and débris.

Under ordinary circumstances, we know that there is a pretty intimate connection between disease and the condition of the weather prevalent at the time, and such an abnormal and decided fall of rain, with its attendant atmospheric conditions generally, cannot but be followed, either directly or indirectly, by some immediate or remote consequences to the health of the community. In previous papers[1] I have endeavoured to show that meteorological conditions such as these are coincident with a low mortality from typhoid fever and an increased prevalence of laryngeal diseases, croup and diphtheria. As we have unfortunately had two other floods in Brisbane within the last few years, it may be of interest to examine the mortality statistics of these two periods in order to test the accuracy of those conclusions, and in the hope that, profiting by past experience, we may have some practical idea of the class of disease that will possibly be more or less common here during the next few months.

A visit to the parts previously inundated would at first lead us to look for an epidemic of either malarial or enteric fever, or both, but in practice this is not found to be the case. It is true that we may meet with a few cases of febricula immediately afterwards, but the symptoms are those *mainly* of congestive or catarrhal conditions, and not of a specific fever, and caused not so much by a specific germ as by direct and prolonged exposure to wet and cold. In one of the papers referred to, it has been stated that typhoid fever is associated with certain atmospheric conditions exactly the reverse of those found in connection with a time of flood—namely, high barometric pressure, high temperature, and high mean and absolute range of temperature, together with low relative humidity, rainfall, and number of rainy days and low amount of cloud. Not only so, but this period is preceded for several months by a similar state of the weather (*see* p. 93), and in the absence of this preparatory period, an epidemic of typhoid fever is probably, unless under special circumstances, an impossibility.

---

[1] *See* "Typhoid Fever," "Diphtheria," and "General Summary."

L

In theory also we should naturally expect malaria to be very prevalent, but we must bear in mind that this is not naturally, or at least has not been for many years, a malarial district, that the conditions favourable to the production of the miasm are not in the subsoil but exist only on the surface, and even there, in the presence of subsequent heavy rainfall, only temporarily. Moreover (*see* p. 114), we should not expect it for several weeks after the subsidence of the flood, so that, upon the whole, the fear of an outbreak of malaria is, from every point of view, somewhat remote. So at least it proved in connection with the two last inundations, for neither at the time nor after the floods in January, 1887, and March, 1890, was there an increased admission rate to hospitals, or mortality, for either malaria or typhoid fever. As far as can be seen, indeed, these floods did not in the slightest degree affect the one or the other in admission or death rate. It is possible that the number of admissions of the nature above indicated may have been increased in the present instance inasmuch as the inundation was greater in degree and continued over a more extended period, but, for reasons already given, this is only a temporary matter and will not materially affect the mortality.

Further examination of statistics for the five years 1887-91, however, shows that the admission and death rate from laryngeal and respiratory diseases rose very high the following autumn and winter respectively.

I. Laryngeal Diseases.—The following table will show the relative magnitude of the mortality from diphtheria for the greater part of the district of Moreton, in April and May, 1887, and in May and June, 1890, as compared to the mean of the other three years under observation :—

MORTALITY from DIPHTHERIA PER 100,000 POPULATION PER ANNUM.

APRIL AND MAY.		MAY AND JUNE.	
1887.	Other Three Years.	1890.	Other Three Years.
121·5	35·5	111·2	69·0

or, in other words, the death-rate from diphtheria in April and May, 1887, was about three and a-half times, and in May and June, 1890, nearly double the mean of the other three years, 1888-89 and 91, for the same periods. The difference here is so decided as to justify us in assuming that there was a pretty intimate connection between the high mortality in both cases, and the previous abnormally high rainfall and generally inundated condition of the town and surrounding district. There can be little doubt that independently of this the mortality would have been specially high in 1890, as the atmospheric conditions favourable to a high death-rate— namely, low barometric pressure, and low mean and absolute range of temperature, high relative humidity and rainfall, and large number of rainy days—were more or less present from the beginning of the year, but that the flood itself materially contributed towards this increased death-rate there can, in my opinion, be little doubt. It is true that in June and July, 1888, the mortality was also high, and that with the exception of a high rainfall in February, and a comparatively large number of rainy days during the first few months of the year, the meteorological observations generally would not lead us to expect a high death-rate. This, however, only shows that an actual flood is not necessary to raise the mortality, and it is possible that had it existed in this instance the rate might have been higher still.

It may here be noted that there was an interval in 1887 of three months, and in 1890 of two months, between the inundation and the period of maximum mortality from diphtheria. Why there should be this difference of one month it is not easy to say, unless it be that in the latter case the flood occurred in March, when the atmospheric conditions are more favourable to the propagation of diphtheria germs than in the hot months of summer.

I have stated elsewhere[1] that the death-rate from diphtheria in the early months of the year depends somewhat upon the state of the weather during the last few months of the preceding year, showing that there is possibly a germination period *in nature* for the development of the bacilli of diphtheria of two or three months' duration, and this assumption is strongly supported by the above considerations.

II. Respiratory Diseases.—As regards these, we find that the admission rate to hospitals for all respiratory diseases in August, 1887, and in May, June, July, and August, 1890, and the death-rate from pneumonia during the months of July, August, and September, are in both cases unusually high. The following figures will show the extent of the admission rate as compared to the mean of the other three years :—

ADMISSION RATE FOR RESPIRATORY DISEASES PER 100,000 POPULATION PER ANNUM.

AUGUST.		MAY, JUNE, JULY, AND AUGUST.	
1887.	Other Years.	1890.	Other Years.
441·6	262·8	435·9	276·0

It will here be observed that the number of admissions to hospitals in this district for the above-noted months in the years of a flood is nearly double the average of other years under observation. The difference in mortality, however, is more decided, as will be seen from the following :—

MORTALITY FROM PNEUMONIA PER 100,000 POPULATION PER ANNUM—JULY, AUGUST, and SEPTEMBER.

1887.	1890.	Other Three Years.
73·6 ...	73·6 ...	17·9

These two tables show that not only was there a greater prevalence of respiratory diseases in the late winter months of 1887 and 1890, but

---

[1] *See* " Diphtheria," pages 14, 19, and 22.

they must have been of a more malignant type, for while the admission
rate for these increased nearly twofold, the mortality from pneumonia,
as will here be seen, increased fully fourfold. This difference in both
admission and death rate is so great that it cannot possibly be a mere
coincidence. It may be mentioned that in both years the winter was
specially cold, the mean temperature in the shade in June, 1887, and July,
1890, being lower by about 5° than that of any other year, and that
the state of the weather otherwise was favourable to a high death-rate
from respiratory diseases. Why that should be so, and whether it
depended in any way upon the previous flood, is a question not for me,
but our able meteorologist, to answer. The flood is at least an
indication of what meteorological conditions may follow and what
diseases we may expect.

On the strength, therefore, of the undoubted connection between
a high relative humidity and rainfall and a high mortality from
laryngeal diseases, supported by experience gained from the last two
floods in Brisbane and district, I would venture to say that we shall probably have an increased prevalence of croup and diphtheria in April and
May, and possibly also of respiratory diseases in the late months of the
coming winter. The latter will, of course, to a great extent depend upon
the meteorological conditions of the next few months, while the former
depends upon two factors that already exist, and probably will exist for
some little time—namely, a suitable soil and favourable weather for
the propagation of the bacilli.

Preventive Measures.—Under these circumstances I hope I may
be pardoned in making a few suggestions, which I must premise are
intended as preventive measures solely for those diseases that will
possibly prevail in the near future, namely, croup, diphtheria,
bronchitis, and pneumonia, and which for the purpose intended may be
followed *only* until their period of maximum intensity—autumn and
winter—have passed over. Should these suggestions be wrong, I hope
you will correct me, and if on the other hand they are incomplete, I
shall expect you to supplement them. In doing so, for laryngeal
diseases these objects have to be kept in view :—

    (A) To remove, destroy, or inhibit the vitality of those
        germs that may already be in process of evolution ;
    (B) To maintain a high state of sanitation, and so provide
        against their future development ;
    (C) To reduce to a minimum the risk of transferring the
        contagium from an infected to a non-infected house or
        district.

In order to carry out these indications, it is advisable :—

    A. 1. That all slush and mud be removed as speedily as possible,
        by whatever means are considered most effectual. Nature
        supplies an efficient scavenger in the rain that falls—
        provided it be sufficiently heavy— subsequent to the
        inundation. In the case of the recent twin-flood, she was
        certainly too liberal in this respect after the first, and
        probably not liberal enough after the second.

2. That some recognised disinfectant be applied freely every day for a week or more to parts in town previously under water, and more especially to swamps near inhabited dwellings.

B. 1. That all damp, marshy ground be efficiently drained. By this means the soil is freed of its superabundant moisture, the relative humidity of the air lowered, and the temperature in winter raised ; or, in other words, the soil as well as the atmospheric conditions are made less favourable for the development of the bacilli of diphtheria. This is a matter that demands *par excellence* the most serious consideration of all, inasmuch as it possibly strikes at one of the most potent factors in connection with the causation of these diseases. The fact of the contagium being directly communicated from one person to another, or from the cow or other animal to man, does not lessen the force of this, or materially increase our knowledge of the original cause of diphtheria; for even though traced in this manner the question still remains, why do these in the first instance suffer from diphtheria, or from some disease that can cause diphtheria in another, and why is it possible for the contagium to be so communicated at one time, and not at another ? We do not know to what extent bogs and marshes may serve as a breeding ground for the development of diphtheria bacilli, but we know that the decaying vegetable matter necessarily existing there in the presence of moisture offers a favourable medium for the propagation of germs generally, that cattle frequently graze on these parts, that the bacilli of diphtheria have been detected in milch cows, and that diphtheria has been communicated from these to man directly through the milk supply. We also know under what meteorological conditions[1] the disease becomes prevalent, that some of these conditions are always present, more or less, near marshy ground, and that the latter, therefore, tends to make the atmospheric conditions more decided, and still more favourable for the development of the diphtheria bacilli.

In connection with the drainage of these parts, it may be well to bear in mind that when from any cause this cannot be satisfactorily carried out they may at least with benefit be converted into ponds. " The conversion of a swamp or low-lying damp piece of ground into a lake will add materially to the dryness and amenity of the climate of the surrounding district ; and the rainier the locality the greater will be the advantage gained.[2]

---

[1] Barometer and mean temperature ... ... low
Relative humidity, number of wet days, and amount of cloud ... ... high.
[2] Buchan.

2. That all public schools, as well as private dwellings, be freely ventilated, and the grounds in connection with these kept absolutely free from rubbish, and as dry as possible. The water supply in both cases to receive special attention and supervision.

3. That all public and private sewers and drains be largely and regularly flushed as often as it is considered necessary by recognised sanitary authorities, and, if required, liberally treated with disinfectants.

4. That all fowl-houses in town be cleaned out every day, and sickly fowls at once killed, removed, and buried.

5. That all mattresses and bedding be exposed as often as practicable to the direct rays of the sun. This is a matter of the first importance, and all the more necessary if the relative humidity of the air be high.

C. 1. That all public dairies be regularly and systematically inspected by a competent health officer or veterinary surgeon, and all eruptive diseases in cattle be notified to the proper authorities, as it has been proved that diphtheria may be communicated not only indirectly but directly from the milch cow to man through the milk supply.[1] It is possible that this may afford some explanation of the undoubted occurrence of diphtheria in houses and districts that are apparently in a perfectly sanitary condition.

2. That all drinking water be boiled and filtered, and all milk boiled, or otherwise cooked, previous to use.

3. That all cases of sore throat be immediately notified and isolated, and should the disease prove to be of a diphtheritic nature, children of the same family prohibited from attending public or private schools. Where isolation cannot be satisfactorily carried out at home, this may be done in a suitable hospital.

I would also venture to suggest that for the next few months cases of sore throat in children should be looked on as possibly in the primary stage of diphtheria, and unless contra-indicated treated by medical men accordingly. No harm is thereby done if the disease proves to be of a less serious nature, and much valuable time saved if our worst suspicions are ultimately realised.

In connection with the possible prevalence of respiratory diseases in winter, besides using ordinary precautions, we may also see that our clothing is of such material as to retard the loss of heat by conduction, and also to some extent by evaporation, according as the weather is wet or dry respectively.[2]

---

[1] Brit. Med. Journal, Nov. 19, Dec. 10, 1892 ; Jan. 7, 1893.

[2] The conditions may thereby be made less favourable for the development of the pneumococci.

www.ingramcontent.com/pod-product-compliance
Lightning Source LLC
Chambersburg PA
CBHW021819190326
41518CB00007B/659